40-

BIBLIOTHECA MYCOLOGICA

BAND 117

Observations on the Bolbitiaceae: 27
Preliminary account of the Bolbitiaceae of New Zealand

by

ROY WATLING & G. MARIE TAYLOR

with 16 figures and 1 plate

J. CRAMER

in der Gebrüder Borntraeger Verlagsbuchhandlung

BERLIN · STUTTGART 1987

Author's addresses:

Dr. Roy Watling
Royal Botanic Garden
Edinburgh EH3 5LR
Great Britain

Dr. G. Marie Taylor
University of Auckland
Dept. of Botany
Private Bag
Auckland, New Zealand

Editores

Prof. Dr. A. Bresinsky, Regensburg; Prof. Dr. H. Butin, Braunschweig und Prof. Dr. H. O. Schwantes, Gießen

© 1987 by Gebrüder Borntraeger, D-1000 Berlin · D-7000 Stuttgart

Printed in Germany by strauss offsetdruck gmbh, 6945 Hirschberg 2
ISBN 3-443-59018-7

CONTENTS

INTRODUCTION

The fungal flora of New Zealand is rather patchily known
in parallel to many countries in Australasia. Whilst many of
the dematiaceous hyphomycetes and their teleomorphs and some
pathogens in New Zealand are currently well documented, and the
gasteromycetes, polypores and resupinate basidiomycetes
relatively well understood (Cunningham 1942; 1963; 1965), the
agaric flora until fairly recently has been very poorly
comprehended.

Some effort was made during the latter part of last century
to catalogue the fungi then recorded for New Zealand (Berkeley,
1855), based primarily on the scores of specimens sent to Kew
notably by missionaries and early settlers such as Colenso, or
explorers such as Berggren; Colenso (1886; 1890) listed many
of his later finds independently. To these records were gradually
added a few more species recognised by their apparent similarity
to well-known European taxa culminating in Massee's publication
(1898) which complemented that of Cooke (1892) on the early
Australian agaric flora. In parallel to many of the
Australian records, however, these compilations are also very
misleading as they too refer to European taxa when in fact it
is now known they should refer to autonomous species. This
discrepancy is undoubtedly because the early workers only had
available texts based on Western Europe and latterly temperate
North American fungi, initiating a tradition which unfortunately
persisted until quite recently.

With such an exciting agaric flora one would have thought of
all the groups of fungi the agarics would have been the first
to have been documented and illustrated in New Zealand, in
parallel to the development of European Mycology. The modern
day student was first made aware of New Zealand's very different
flora by a series of papers by Stevenson-Cone (Stevenson
1962a, b & c; 1964). This was extended still further by the
late Ross McNabb (1967; 1968; 1971; 1972; 1973), and then
by the analysis of particular genera by Horak (1971a,

1973a, b, c, d & e; 1977; 1979a; 1980a, b & c) and Heinemann
(1974). In addition to these contributions Horak published an
analysis of many of the early collections housed in national
herbaria (Horak 1971b & c) subsequently using this as a base-
line for further work. Thus many of Colenso's collections were
located and re-examined and type material documented; some of
Colenso's collections not in Royal Botanic Gardens, Kew (K) and
therefore not analysed by Horak were subsequently located by
the present authors in Edinburgh (E). Surprisingly relatively
few popular illustrated accounts on higher fungi have been
produced to date (Taylor, 1968; 1970; 1981; 1983; Stevenson,
1982a; Bell, 1983).

The present contribution is a continuation of all this work
and is offered as a preliminary documentation of the New Zealand
members of a family not as yet treated, i.e. Bolbitiaceae. Many
of the constituents of this family are so dull-coloured and
insignificant, and often small, that they have not been
collected in any great number in the past. This paper is based
largely on collections made over a ten year period by one of us
(GMT) and by Drs Barbara Segedin (Dept. of Botany, Auckland) and
P. Austwick (Brompton Hospital, London). Egon Horak, Zürich,
has been kind enough to communicate his collections of the
Bolbitiaceae from Australasia and Professor E.J.H. Corner,
Cambridge, England has also placed his material at our disposal.

The following Colenso and Berggren collections have been
examined; those not located by Horak (1971b) are indicated by
asterisk.

Colenso	b	71 Agaricus vervacti	Berggren 60 Galera tenera
		- Agaricus erebius	61 Agaricus semi-
	b	113 Agaricus pediades*	orbicularis
		269 Agaricus pediades	
	b	283 Agaricus praecox	
	b	874 Agaricus pudicus*	
		1053 Agaricus semiorbicularis*	
		- Agaricus strophosus	

Summary of New Zealand taxa:

Agrocybe Fayod

Subgenus _Agrocybe_ Sect. _Agrocybe_

A. acericola (Peck) Singer

A. howeana (Peck) Singer

A. praecox (Persoon: Fries) Fayod

A. puiggari (Speg.) Singer

Agrocybe sp. 1

?_Agrocybe_ sp. 2

Sect. _Pediadae_

A. olivacea Watling & Taylor

A. pediades (Fries) Fayod

A. semiorbicularis (Bulliard: St Amans) Fayod

A. temulenta (Fries) Singer

Sect. _Microsporae_ fide Singer

A. vervacti (Fries) Singer

Subgenus _Aporus_   Sect. _Velatae_

A. erebia (Fries) Kühner

Sect. _Aporus_

A. parasitica Stevenson

Agrogaster Reid

A. coneae Reid

Bolbitius Fr.

Subgenus _Pluteolus_

B. muscicola (Stevenson) Watling

Subgenus _Bolbitius_

B. titubans (Persoon: Fries) Fries

B. vitellinus (Persoon: Fries) Fries

Bolbitius spp. (1 & 2)

Conocybe Fayod

Subgenus _Conocybe_ Sect. _Conocybe_ (Farinosae)

C. mesospora Kühner ex Kühner & Watling

C. pubescens (Gillet) Kühner

Conocybe spp. (1-4)

Sect. _Pilosellae_

<u>C</u>. aff. _piloselloides_ Watling

Sect. _Candidae_

C. huijsmanii Watling

Subgenus _Ochromarasmius_

C. _horakii_ Watling & Taylor

Subgenus _Piliferae_

C. _novae-zelandiae_ Watling & Taylor

Subgenus _Pholiotina_

C. _gracilenta_ Watling & Taylor

<u>C</u>. _rugosa_ (Peck) Watling

<u>C</u>. aff. _vexans_ P D Orton

Systematics

This contribution covers three genera, _Agrocybe_, _Bolbitius_ and _Conocybe_ as defined by Watling (1965; 1982) and for completeness _Descolea_ which, although formerly placed in the Cortinariaceae by Singer (1951; 19620, has been transferred by him to the Bolbitiaceae (Singer, 1969). We agree with Horak (1971a) that _Descolea_ should be maintained in the Cortinariaceae.

The genus _Agrogaster_ with a single species _A. coneae_ was described by Reid 1986. This secotioid to agaricoid species was placed in the Bolbitiaceae near to _Agrocybe_ for the following reasons: the cap pellis is a palisadoderm, cheilo- and pleurocystiolia with swollen base and tapered apex are abundant and the basidiospores are pale brown.

The vegetation of New Zealand has been considerably modified by man over the generations especially over the last hundred and fifty years, and many of the taxa listed below characterise disturbed land. The flora undoubtedly contains both native, endemic species and introduced taxa, the last very probably connected with farming practices. The endemics will not be recognised until similar studies are completed in neighbouring countries.

BOLBITIACEAE

Key to Genera

1. Gills free or adnate becoming free; pileus viscid, thin and
    soon decomposing, appearing as if deliquescent.
    Basidiomata brightly coloured; purplish, pinkish
    or yellow, less frequently lacking colour.
    Growing on wood, soil, dung or herbaceous debris 2. <u>Bolbitius</u>

1x Gills rarely free and, if so, then pileus not
    viscid, nor thin and 'deliquescent'. Basidiomata
    usually some shade of brown.   2

2. Spore-print rust- or cinnamon-colour   3. <u>Conocybe</u>

2x Spore-print hazel, snuff- or cigar-brown   1. <u>Agrocybe</u>
    Colours whenever possible follow Henderson, Orton and Watling
    (1969), an Identification Colour Chart, Flora of British Fungi.

AGROCYBE

    <u>Basidiomata</u> medium to large, only occasionally small,
generally dull-coloured or pallid to cream-coloured, fleshy,
collybioid or sometimes tricholomatoid, never deliquescent.
<u>Pileus</u> often comparatively fleshy, convex to semiglobate then
expanding, humid, tacky or dry, smooth or slightly wrinkled,
not plicate-striate, uncommonly markedly striate, appendiculate
velar remnants present at margin in some species. <u>Stipe</u> central,
white, buff or ochraceous, darkening from base-up, usually
attached to long, white mycelial cords, with distinct ring or
velar fragments, or veil absent. <u>Gills</u> adnate to adnexed or
even subdecurrent in one group of species, pallid to ivory then
buff or ochraceous but soon becoming hazel, snuff- or cigar-brown,
usually with floccose-flocculose margin. <u>Flesh</u> white or pale-
coloured, in stipe generally darkening upwards with age; taste
and smell often mealy (farinaceous).

    <u>Basidiospores</u> smooth, generally with thick wall and prominent
germ-pore. <u>basidia</u> usually 4-spored, in a few species 2-spored;
<u>gill trama</u> regular becoming less regular with age. <u>Pileipellis</u>
a distinct palisadoderm at first with or without dermatocystidia
but may soon disorganize and become mixed with filamentous
units; 'scalp' cellular. <u>Cheilocystidia</u> always present,
vesiculose to lageniform or cylindric; <u>pleurocystidia</u> present
or absent, either simple or prominently differentiated.

Stipitipellis of cylindric, hyaline to slightly brownish hyphae;
caulocystidia usually present especially at stipe-apex.

Development where known paravelangiocarpic, except in one
species (A. cylindrica) where it is bivelangiocarpic.

On the ground in woods, pastures, heaths etc.; on dung,
plant remains, refuse and wood; often in gardens, farm-yards
and greenhouses; saprophytes, or in the case of A. parasitica
a weak parasite.

Key to Sections

1. Veil absent   2
1x Veil forming either a ring on the stipe or an
    appendiculate veil to pileus-margin   5

2. Cystidia numerous on pileus, stipe and gills   3
2x Dermatocystidia absent or if present inconspicuous;
    pleurocystidia absent or if present never highly
    differentiated   4

3. Cystidia thick-walled, glassy or with silvery
    appearance   Allocystides
                                    (Not represented in New Zealand)
3x Cystidia never thick-walled except occasionally in
    pedicel; basidiospores not necessarily small but
    often with small germ-pore   Microsporae
                                    (Not represented in New Zealand)

4. Pleurocystidia sparse or absent, or if present
    then basidiospores < 11.5 µm long with prominent
    germ-pore   Pediadae
4x Pleurocystidia present; basidiospores ⩽ 8 µm long
    and with reduced or absent germ-pore   Evelatae

5. Pleurocystidia usually numerous often quite large;
    basidiospores with broad truncate germ-pore   Agrocybe
5x Pleurocystidia few and often overlooked or if
    present basidiospores either elongate (almost
    boletoid) and/or with reduced or absent germ-pore   6

6. Basidiomata when fresh sepia, cigar brown etc;
   basidiospores elongate-subfusiform often with
   apical papilla; terrestrial   Velatae

6x Basidiomata when fresh fulvous to brick then
   ochraceous; basidiospores ellipsoid; growing
   on dead or living trees   Aporus

Key to Species

Taxa not recorded for New Zealand but which might be
expected there are in parenthesis.

1. On wood, growing from wounds at base of trunk
   or higher; spores 8-12 x 5.5-6.5 µm; germ-
   pore reduced or even absent   4. _A. parasitica_

1x Growing on soil   2

2. Stipe annulate   3

2x Stipe exannulate (if with roughened fibrillose
   stipe with velar remains, see 12 _Agrocybe_
   sp. 1)   8

3. Chrysocystidia present on gill-face; spore-
   print purplish brown; pileus-scalp
   filamentous   _Stropharia coronilla_ see pgs 26 & 55

3x Chrysocystidia absent; 'scalp' cellular or a
   mixture of rounded cells and a few
   filamentous hyphae (see 13 _Agrocybe_ sp. 2)   4

4. Pleurocystidia digitate, with long finger-like
   processes   See notes to
                             1. _A. acericola_ & 2. _A. praecox_

4x Pleurocystidia either mucronate to ampulliform
   or utriform to vesiculose   5

5. pileus dark-coloured; basidiospores (7-)7.5-8.5(-10)
   x 5.5-6.5 µm   13. _Agrocybe_ sp. 2

5x Pileus buff, tan or cream-coloured;
   basidiospores usually >8.5 µm   6

6. Cheilocystidia inflated, vesiculose,
    sometimes with short neck   7

6x Cheilocystidia narrow, lageniform to
    elongate-cylindric   8. <u>A. puiggarii</u>

7. Pleurocystidia mucronate to ampulliform   6. <u>A. praecox</u>
                                               (also see 4. above)

7x Pleurocystidia inflated utriform to
    vesiculose   3. <u>A. howeana</u>
                                               (also see 4. above)

8. Pileus olivaceous; stipe with zones
    of veil   10. <u>A. olivacea</u>

8x Pileus ochraceous; stipe not zoned   9

9. Basidiospores 7.5–8.5(–9) μm long   11. <u>A. vervacti</u>

9x Basidiospores longer, over 10 μm   10

10. Basidiospores 14–16 x (8–)8.5–11 μm; basidia
    2-spored   9. <u>A. semiorbicularis</u>

10x Basidiospores (10.5–)11.5–13(–14) x 7.5–8(–8.5)
    μm; basidia 4-spored   5. <u>A. pediades</u>

1. <u>A. acericola</u> (Peck) Singer in Acta Inst. Bot. Komarov Acad.
Sci. URSS., Series 2; 6: 88, 1950. Fig. 1 K–N.
Basionym: <u>Agaricus</u> <u>acericolus</u> Peck in Bull. Buff. Soc. Nat. Sci.
1: 50, 1873.
Synonym: <u>Pholiota acericola</u> (Peck) Sacc. in Sylloge Fungorum 5:
759, 1887.

    A single collection from New Zealand possibly could be
placed here. Unfortunately this collection from Mt Albert,
Auckland in <u>PDD</u> (K.R.W. Hammett, 29208) is supported only by

scanty field data, viz. (Spores cinnamon brown. Pileus cream,
white, smooth, adnate* gills. Stem fibrillose, hollow).

The microscopic data of this collection are: <u>Basidiospores</u>
(7.5)8-9 (-9.5) x 4.5-6 µm, ellipsoid, very slightly amygdaliform
in side-view, smooth, relatively thick-walled; germ-pore
distinct. <u>Basidia</u> 4-spored, 20-22.5 x 5.6 µm. <u>Cheilocystidia</u>
collapsed, clavate to vesiculose, 12.5-17.5 x 8-8.5 µm;
<u>pleurocystidia</u> not prominent, ventricose with papillate apex
or digitate appendages, 35-40 x 13-20.5 µm, apex 3-6 µm broad.
<u>Pileipellis</u> a palisadoderm of spheropedunculate cells 15-25.5 µm
broad. <u>Gill-margin</u> matted with filamentous flexuous cells
40 µm long resembling those seen in <u>A. parasitica</u> q.v.

<u>Segedin</u> 1631 (Mt Eden, Auckland, legit Hasnain, 19 ix 1979)
had similar basidiospores and cheilocystidia and some
pleurocystidia with digitate apices similar both to those
figured by Overholts (1927) for <u>A. acericola</u>, and seen in
collections by one of us (RW) of the same species from Michigan,
USA.

This fungus was originally described as <u>Agaricus</u> (<u>Pholiota</u>)
<u>acericolus</u> Peck from wood and trunks of maple (<u>Acer</u>) covered
with moss but it is not infrequent in N. America on rotten <u>Acer</u>
spp., old trunks of <u>Fagus grandifolia</u> Ehrh. and on piles of
sawdust and chips from felling activities in frondose woodland
or used as mulching in shrubberies. The Auckland material was
on wood-chips and could conceivably have been introduced into
New Zealand.

2. <u>A. erebia</u> (Fries) Kühner, Le Genre Galera, 31, 1935.
Basionym: <u>Agaricus erebius</u> Fries, Systema Mycologicum I: 246,
1821.
Synonym: <u>Pholiota erebia</u> (Fries) Gillet, Hyménomycètes de France
6, 86, 1874.

This was listed by Berkeley in Hooker's New Zealand Flora
(1855) and Massee (1898) as <u>Agaricus (Pholiota) erebius</u>

(* Crossed out)

collected by Colenso from Ahuriri; the record should be
accepted with some reservation, although it may indicate that
this taxon is present in the New Zealand flora. Cunningham is
purported to have collected this taxon at Weraroa (material in
PDD) but this is a quite different fungus, and as much as we
know about the collection is tabulated below; see pg 28.
A. erebia should be looked out for under shrubs in gardens and
parks, and if present is probably introduced.

The microscopic characters which delimit A. erebia in Europe
are:-

Basidiospores 9-13(-15) x (5-)6-7 µm, elongate elliptic in
face-view, slightly boletoid in side-view, ochraceous in water
and alkali; germ-pore absent. Basidia 2-spored, cylindric-
clavate, hyaline. Cheilocystidia variable, vesiculose to
ventricose even clavate, 30-60 x 8-25 µm, hyaline; pleurocystidia
few to numerous, lageniform, 50-80 x 12-16 µm, apex obtuse, (4-)
6-9 µm broad, hyaline often granular punctate. Pileipellis a
palisadoderm of balloon-shaped to pyriform cells 9-15 µm broad.
Stipitipellis of filamentous hyphae supporting at stipe-apex
sterile cells similar to the cheilocystidia. Clamp-connections
apparently absent.

Another closely related species but in too poor condition to
determine occurred in a garden at Weraroa, see page 27.

3. A. howeana (Peck) Singer in Lilloa 22 (1949): 492 (1951).
Fig. 2 A,B,D & E.
Basionym: Agaricus (Stropharia) howeanus Peck in Bull. Buff. Soc.
Nat. Sci. 1: 53, 1873.
Synonym: Pholiota howeana (Peck) Peck in Bull. New York State
Mus. Nat. Hist. 122: 147. 1908.

Description of New Zealand material:
Pileus 25-40-100 mm, hemispherical then plano-convex to plane
or slightly gibbous, near pale luteous, then brownish cream,
darker towards centre, smooth, dry, becoming cracked,

hygrophanous; margin serrate. <u>Stipe</u> (30)50-100 x 3.5-15 mm,
equal, ± curved, fragile, annulate, slightly lighter than pileus
or buff, darkening below, stuffed then hollow; ring darker buff,
superior, fragmentary, when persistent fibrous. <u>Gills</u> free or
slightly adnexed, adnate-sinuate, ± crowded, fawn, darker with
age to become near umber. <u>Flesh</u> white, thick under pileus-disc.
<u>Taste</u> mild; <u>smell</u> disagreeable.

<u>Basidiospores</u> (8-)8.5-9.5 x (5-)6-6.5(-7) μm, ellipsoid-ovoid
with smooth, relatively thick wall, brown in both water and
alkali; germ-pore present. <u>Basidia</u> 4-spored, 20-23 x 5.5-7.5
μm, clavate. <u>Cheilocystidia</u> mixtures of clavate and ventricose
cells, sometimes utriform, 25-38 x 15-20 μm; <u>pleurocystidia</u>
prominent, relatively numerous, ventricose, (25-)40-50 x
15-18 μm. <u>Pileipellis</u> a palisadoderm of spheropedunculate cells
13-20.5 μm broad.

<u>Material examined</u>: Gregarious on lawn, Mt Roskill, Auckland,
19 v 1971, legit K.R.W. Hammett. <u>PDD</u> 29104.

As presently described by Singer (1978) this is a rather
broadly conceived taxon and probably will be divided into at
least two taxa in the future. As the New Zealand material does
not allow detailed examination the above epithet is utilised
meantime; the New Zealand material was said to have a mild not a
bitter taste, usually a character of <u>A. howeana.</u> It is found
not only in North America, from where Peck described it, but
also S. America and Central Africa, where it was probably
introduced. It should be noted that Peck firstly referred this
species to <u>Stropharia</u> because of the purplish tinge to the
immature gills; comparison should be made with <u>Stropharia</u>
<u>coronilla</u> pg. 7. A second collection in <u>PDD</u> (solitary,
amongst grass on ground, Weraroa, Wellington, 2 x 1919,
Cunningham 680) is hesitantly placed here; the pleurocystidia
are not as frequent. Some of Cunningham's information is
utilised in the description above but because the species is
so poorly documented in New Zealand a description of N. American
material is offered for comparison:-

Pileus 53-100 mm, convex then expanding, becoming wavy at margin
in broad undulations, rich tawny (pale)-straw colour paling to
straw yellow at concentrically wrinkled margin, darker but often
only slightly so towards disc or in larger basidiomata tawny
orange at disc and similar although pale outwards, humid, smooth
or fairly rugulose. (One basidioma with pileus tinged distinct
olivaceous buff throughout) Stipe 40-68 x 8-13 mm, narrow and
relatively thin for size of pileus, swollen slightly at stipe-
base to form small bulb with white rhizoidal strands, annulate,
pale buff to pallid, irregularly fibrillose-striate below ring,
striate at white apex, hollow; ring apical, thin, membranous,
soon collapsing onto stipe and then almost disappearing or
leaving in age a thin skin from collapsing or after abrasion.
Gills buff-pallid with a purple-brown tinge especially when
young then becoming snuff-brown. Flesh white in pileus,
pinkish brown tinge in stipe downwards, within stipe around
'mid' brown. Taste bitter-astringent not mealy (or at most
only a hint which is masked and not persistent); smell not
mealy, unpleasant but not persistent or penetrating, very
earthy, of vegetables immediately on cutting.

In clump; under collapsed Manzanita (Arctostaphylos sp.)
by trackside, Selyer, near Willow Creek, N. California 11 iii
1984, Wat. 17222 in E.

4. A. olivacea Watling & Taylor, sp. nov. Fig. 3 A-E & N
Pileus <52 mm plano-convexus vel subumbonatus, brunneo-
olivaceus, centraliter ochraceus et rugosus. Stipes 150 x 6 mm
(13 mm ad basem) bubalinus, ad apicem attenuatus et gradatim
pallescens, squamulis ochraceis subcingulatis irregulariter
basim versus instructus, tandem deorsum fuscans, rhizomorphis
albis crassis longisque praeditus. Lamellae porphyrobrunneae,
subliberae. Caro sapore miti.

Basidiosporae cumulatae brunneae, 11-12.5 x 7-8 µm
ellipsoideae, laeves, tunica crassa poro germinativo conspicuo
instructae. Basidia 4 -sporigera, 34-40 x 10-11 µm, clavata.

Cystidia aciei lamellarum vesiculosa, catenata, fibulata,
20-30 x 15-18 µm. Cystidia faciei lamellarum clavata apice
dilatato 46-62 x 23-28 µm. Cystidia stipitis lageniformia.

Ad terram inter ramos mortuos, Novazelandia. Typus Horak
ZT 69/88.

Pileus <52 mm, convex at first, becoming umbonate-expanded,
honey-brown to ochre brown at disc, pale grey-brown outwards,
centre distinctly wrinkled to grooved, less so towards margin,
hygrophanous, glutinous; margin inconspicuously striate.
Gills adnexed to almost free, very dense, grey with distinct
lilac tinge, with concolorous, subserrate to fimbriate edge.
Stipe 150 x 6 mm (13 mm at base) cylindrical or attenuated
towards apex, base also subbulbous, with conspicuous white
rhizoids, whitish to pale yellowish, dry, hollow, minutely pruinose
at apex, below with numerous appressed concolorous to pale
yellow-brown belts of veil; cortina none. Chemistry KOH,
HCl, NH$_3$ on pileus-negative.

Basidiospores brown in mass, 11-12.5 x 7-8 µm, elliptic
in face-view, slightly flattened in side-view, thick-walled,
smooth; germ-pore relatively large. Basidia 4-spored, 34-40 x
10-11 µm clavate. Cheilocystidia vesiculose often in chains
separated by clamp-connections, end-cells 20-30 x 15-18 µm;
pleurocystidia clavate with swollen apex.

Material examined: Westland, Harihari, Big Wanganui, on
soil amongst debris of Dacrydium, Hedycarya, Weinmannia etc,
17 ii 69; Horak ZT 69/88.

This is a member of Agrocybe sg. Agrocybe sect. Agrocybe
in virtue of the prominent pleurocystidia and velar remains. It
undoubtedly is related to A. earlei (Murr.) Watl. from which
it differs in slightly broader spores, more vesiculose
pleurocystidia and olive pileus-colour.

5. A. parasitica Stevenson in New Zealand Journ. of Forestry
27: 130, 1982. Figs 4 & 6.
Synonyms: Agrocybe cylindracea s. Taylor, 1981.
           Agaricus (Pholiota) pudicus s. Colenso, 1890 and
           Massee 1898.

Pileus 65-200 mm, convex to plano-convex finally plane, at
first dark yellow brown to greyish brown ("snuff brown" "fulvous")
then fading during expansion to buff with cinnamon tinge at disc,
at length wholly whitish, smooth with a texture of close velvet
becoming like chamois then naked on weathering, occasionally
dimpled in age, often becoming cracked into areolae; margin
entire, not striate, dotted with loose crumb-like particles
whilst still inrolled. Stipe 110-150(-200) x 12-20 mm,
cylindric to somewhat thicker downwards, annulate, white to
pale buff with longitudinal lines of concolorous to brownish
tiny squamules, surface occasionally cracking into recurved
scales downwards; ring apical, membranous, pale buff, slow to
detach from pileus-margin, collapsing into a full "flared
skirt", ridged and striped above by longitudinal dark brown
lines from spores and sprinkled below with crumb-like fragments
similar to those on the young pileus margin, soft and collapsing
in humid weather. Gills adnate with slight tooth, fairly
crowded from many lamellules, buff darkening to snuff-brown or
umber, margin white, finely floccose-denticulate. Flesh white
in pileus and stipe-apex, firm, a grey watery zone sometimes
present above gills, browning towards stipe-base, tough-fibrous
in stipe-cortex, more tenuous inside. Taste mild, pleasant;
smell nutty, like yeast.
Basidiospores umber to date-brown in mass, (8.5-)9.5 -12 x
(5.5-) 6-6.5 µm, ellipsoid, slightly thick-walled, ochraceous
brown in water, darker in aqueous alkali solutions, guttulate,
non-amyloid, not strongly cyanophilic; germ-pore present,
small. Basidia (2-)4-spored, clavate-cylindric. Cheilocystidia
lageniform swollen below with short neck, to ampulliform,
intermixed with ventricose cells, 18-35 x (7-)9-10 µm, hyaline
or slightly honey-coloured in water and alkali solutions, in two
collections masked by velar remnants; pleurocystidia present,
prominent, fairly numerous, similar to cheilocystidia although
neck sometimes longer (>6 µm), 24-40 x 6-11 µm, neck 3-5 µm

broad. <u>Hymenophoral trama</u> regular, incolorous. <u>Pileus trama</u>
with oleiferous hyphae; clamp-connections numerous. <u>Pileipellis</u>
a palisadoderm through which hyphae 5-6 μm broad project,
sometimes accompanied by cystidioid cells. <u>Velar remnants</u> at
margin of pileus and edge of gills hyphoid, clamped, end-cells
28-36(-70) x 3.5-6 μm, intermixed with cystidia. <u>Clamp
connections</u> present.

<u>Habitat</u>: gigantic clusters or singly, on introduced and native
dicotyledonous trees especially tawa (<u>Beilschmiedia tawa</u> Benth.
& Hook. f.) and kohekohe (<u>Dysoxylum spectabile</u> Hook. f.);
common, widespread and almost certainly indigenous in New
Zealand. It has not as yet been recorded on any gymnospermous
host. It is found both on living hosts and those recently
dead and sometimes fairly high above the ground. Basidiomata
appear typically from wounds, being associated with a very
slow heart rot, although it is probably only a secondary invader;
but see Stevenson, 1982.

This species is common throughout New Zealand and probably
recognised by Colenso (1890) and Massee (1898) as <u>Agaricus
(Pholiota)</u> <u>pudicus</u>. Indeed Colenso's record of <u>Pholiota
heteroclita</u> (Fr.) Quél. may refer to the same species but the
material (<u>b</u> 624) in K lacks basidiospores. It is an excellent
edible mushroom very closely related to <u>A. cylindrica</u> (DC.:Fr.)
Maire, indeed at the start of the present study following Singer
(1969) the present authors included it within their circum-
scription of <u>A. cylindrica</u>.

<u>Material examined</u>: Colenso <u>b</u> 874, as <u>Agaricus pudicus</u>
(K); occasionally single, or 2 or 3 together, growing from
trunk of living tree in semi-swampy forest with tawa
(<u>Beilschmiedia tawa</u>) and kahikatea (<u>Podocarpus dacrydiodes</u>
A. Rich.), Wainui-o-mata Valley, at Catchpool Stream, 25 iii
1961, <u>Taylor</u> 50 & 52; on standing <u>Weinmannia racemosa</u> Linn. f.,
Tawa track, Urewera Nat. Park, 1 iii 1970, <u>Austwick</u> 1961; on
trunk of living tree, Gudex Park, Maungakawa, near Cambridge,
9 iv 1970, legit E P White, <u>Austwick</u> 2042; on trunk of tree,
Ruakura Research Centre, Hamilton, 28 iv 1970, <u>Austwick</u> 2124;

on _Leptospermum ericoides_ A. Rich., Dingle Dell, Auckland, 22 iii
1972, _Taylor_ 702; in cleft at base of dead tawa trunk (_B. tawa_),
Cossey's Dam, Hunua Range, 17 xi 1973, _Taylor_ 867; on
Metrosideros excelsa Soland. ex Gaertn., St Andrews road, Epsom,
Auckland, 21 iii 1979, _Segedin_ 1564; on unidentified wood,
Clive Road, Epsom, Auckland, 3 xii 1980, _Segedin_ 1723; on dead
tawa (_B. tawa_), in native forest, volcanic plateau, Whirinaki,
13 i 1981, _Segedin_ 1732; on _Rhododendron_, St Andrews Road,
Auckland, 6 ii 1981, _Segedin_ 1734; on rotten wood, Piha, Waitakere
Range, Auckland, 5 v 1981, legit P. Johnston, _Horak_ ZT 503; on
living standing tawa (_B. tawa_), Waikoha Road, Whatawhata, legit
Heather Curran, 27 xii 1981, _Taylor_ 1157; on _Erythrina caffra_
Thunb., Old Government House, University of Auckland, legit A.
Palmer, 27 iii 1982, _Taylor_ 1180; on _E. caffra_, as above, 5 iv
1982, _Taylor_ 1188; Hakarimata Walkway, Waikato, 19 xi 1983,
Taylor 1365; on _Macropiper excelsum_ Miq., in cleft 0.5 m above
ground level, Therkelson Walk, New Plymouth, 27 xii 1983,
Taylor 1370; on tawa (_B. tawa_), Ratapihipihi Reserve, New
Plymouth, 1 i 1984, _Taylor_ 1374.

In addition Stevenson (1982 a & b) describes the same
species on _Alectryon excelsum_ Gaertn., _Carpodetus serratus_
Forst., _Corynocarpus laevigatus_ Forst., _Dysoxylum spectabile_
Blume, _Hoheria_ sp., _Laurelia novaezelandiae_ A. Cunn., _Nothofagus
solandri_ Oerst. and _Plagianthus betulinus_ A. Cunn. To this
list can now be added _Erythrina caffra_, _Leptospermum ericoides,_
Macropiper excelsum, _Magnolia grandiflora_ Linn. (Univ. of
Auckland), _Metrosideros excelsa_, _Populus italica_ (Duroi) Muench.
(Wynyard Car Park, Univ. of Auckland), _Rhododendron_ and _Senecio
grandifolius_ Berg.

Thus _A. parasitica_ has now been found on both native and
introduced trees and would appear to be far commoner than
A. cylindrica is in Europe which, although it has been recorded
on several hosts, is much more restricted in its host preference.
One record exists in New Zealand for _A. parasitica_ on _Ulmus glabra_
Huds., a host on which _A. cylindrica_ is found in Europe.
Illustrations of this species occur in Taylor (1981; 1983).

A. parasitica (as _A. cylindrica_) from Australia is

illustrated by Macdonald and Westerman 1979 and Fuhrer 1985. Dr
Fuhrer kindly sent a collection to Edinburgh for study and it
confirmed our identification. Material examined: on _Atherospermum
moschatum_, Talbot Drive, near Marysville, Vic., Australia, v 1985,
B. Fuhrer B236. This may be the first record of _A. parasitica_
in Australia, where it appears to be not uncommon in Victoria
at least, and is probably widespread.

Singer (1969) synonymised _Agaricus crassivelus_ Spegazzini
(in Ann. Soc. Cienc. Argent. 9: 279, 1880), _Pholiota formosa_
Spegazzini (in Bot. Acad. Cienc. Cordoba 28: 311, 1926) and
Pholiota impudica Spegazzini (in Bol. Acad. Nac. Cienc. Cordoba
11: 414, 1889) with _A. cylindrica_ (as _A. aegerita_) and recognised
considerable variation within this taxon. He suggested that
several taxa may ultimately be described in the complex.
A. parasitica is just one such taxon; unfortunately during the
present study Spegazzini's types were not examined and so it has
not been possible to compare the microscopic characters
demonstrated in our material with those published for the
S. American taxa. However, collections from Brazil examined by
one of us (RW) demonstrate that quite a different fungus is at
least present there. _Agaricus phylicigenus_ Berkeley from Tristan
da Cunha is also claimed to be _A. cylindrica_ (Singer, 1969),
but the type in K is so badly preserved that it does not help
to clarify the situation.

We have compared our material with the following collections
of _A. cylindrica_ from Europe (all in E) – The Netherlands: on
Populus. Italy: on _Ficus_, Genova, Erbar Crittogam. Ital., Ser.
II. British Isles: Petersham, Surrey, 27 v 1908, legit Hartley
Smith; on _Populus_, Hereford, Ross-on-Wye, 29 x 1969, _Orton_ 3503;
on _Acer campestre_ L., Archer Wood, Coppingford, Huntingdonshire,
5 v 1973, legit Sheila S Wells; on _Populus_, Offord,
Huntingdonshire, 7 vii 1974, legit S. Wells; on _Sambucus_, Castor
Henglands NNR, Cambridgeshire, 3 x 1976, legit S. Wells. Other
material from the British Isles was examined during the preparation
of the British Fungus Flora (Watling, 1982) and the data obtained
therefrom has been used in this study.

In the Dutch and some English material on _Populus_ the

pleurocystidia are very rare, and the cheilocystidia vesiculose
to ventricose and never with a long appendage. This is in fact
parallel to material collected in Kashmir (on <u>Populus alba</u> L.,
Gulmarg, 28 ix 1978, <u>Wat</u>. 13067 in E & Abraham - <u>6RRL</u> 7748 in E);
see Watling & Gregory, 1980 and Watling & Abraham (ined.). In
those British collections with abundant pleurocystidia these are
always clavate, although often with a long pedicellate base,
and subhymenial or even tramal in origin. In this way they
resemble collections from Afghanistan (in garden, Kabul, 12 v
1951 legit Gehli; on <u>Populus pyramidalis</u> Rozier (= <u>P. italica</u>
(Duroi) Muench) Kabul, 21 v 1951; both det. R. Singer in MICH),
although the basidiospores of these agree more with <u>A. parasitica</u>
(11-12 x 6-7 μm).

Thus whereas the European collections appear to be hetero-
geneous, those from New Zealand are uniform and characterised by
the following cluster of characters:- relatively large
basidiospores with small, although quite distinct germ-pore,
elongate cheilocystidia and abundant development of pleurocystidia
with long apical appendage.

<u>Taylor</u> 867 is notable in showing an excellent development of
hyphal material from its bivelangiocarpic growth in parallel
to the condition described for <u>A. cylindrica</u> by Reijnders (1979);
however, this collection consists of immature basidiomes, judging
from lack of spore-development; the basidiospores available
are in fact comparatively small.

Some variation is found in the detailed structure of the
pilepellis in the collections of <u>A. parasitica</u> but this simply
reflects the state of maturity. At first the pileus is covered
in hyphal strands under which a palisadoderm is formed; as the
pileus expands these hyphae become intermixed with ellipsoid-
pedicellate to spheropedunculate cells and at maturity are
even joined by hyphae derived from the pileus-trama. This may
explain the variation found in <u>A. cylindrica</u> where rugulose
and smooth-capped forms occur but more critical notes are
required on individual collections to ascertain the significance
of such a character and whether it is correlated with other
micro-data.

The variation in basidia, cystidial morphology and their

abundance in Brazilian collections examined by one of us (RW)
indicates that almost totally 2-spored forms are common there,
2- and 4-spored basidia were found on many of the New Zealand
collections, e.g. Taylor 867, and in European collections from
the Netherlands and from England (Orton 3503). Singer (1969)
indicates that similar variation is found in the S. American
collections he has examined, he also indicates that within
these same collections of A. cylindrica (as A. aegerita) he
found populations in which the basidiospores are distinctly
truncate whereas in others they are not, populations in which
the the pleurocystidia are acute and others in which the
pleurocystidia are obtuse ventricose-ampullaceous, and
populations which are exclusively 2-spored and others exclusively
4-spored.

As the basidiospores of members of the A. cylindrica group
germinate so readily and the axenic cultures prepared fruit
readily these agarics offer ideal subjects for the study of
variation and interfertility. Esser and his colleagues (Esser,
Semerdzieva & Stahl, 1974; Esser & Meinhardt, 1977; Meinhardt
& Esser, 1981) have demonstrated the ways by which variability
in European isolates can be studied, indeed they have analysed
some 2-spored forms. Thus the conspecificity or autonomous
nature of the various isolates in the A. cylindrica complex
could be determined in parallel to the recent techniques used
for analysing ranges of Armillaria isolates. A. cylindrica
is in fact cultivated commercially in Europe; see Ferri, 1973.

6. A. pediades (Fries) Fayod in Ann. Sci. Nat. (Bot.) Series 7,
9: 358, 1889. Fig. 7 A-K.
Basionym: Agaricus (Psilocybe) pediades Fries, Systema
Mycologicum I: 290, 1821.
Synonym: Naucoria pediades (Fries) Kummer, Die Führer in die
Pilzkunde 22: 78, 1871.

The New Zealand material possessed the following characters:
Pileus hemispherical convex or plano-convex, pale tan or
golden tan, smooth. Stipe equal, neither ring nor velar
fragments present. Gills adnate-sinuate, blackish brown.

Basidiospores 12-14 x 7.5-8.5 µm, elliptic in face-view, slightly flattened in side-view, thick-walled; germ-pore broad, central. Basidia 4-spored, 33-34 x 9-11 µm, cylindrico-clavate with basal clamp-connections, hyaline, some pale yellowish. Cheilocystidia 19-35(-40) x 5-12 µm tibiiform, with subcapitate to capitate head 2-5 µm broad, some with tapered neck, non-capitate 35-48 x 5-10 µm with apex 2-3 µm broad; pleurocystidia absent. Caulocystidia supported by cylindric hyphae with clamp-connections, in shape similar to those on gill-edge, varying from cells with tapered neck to cells with or without capitate head, some with pale yellow incrustations in alkali solutions. Clamp-connections present.

Material examined: on sandy roadside, disturbed by road-grader, Skipper's Canyon, 28 xi 1982, Taylor 1263; on soil, Brian Duder's farm paddocks, Clevedon, 16 iv 1983, Taylor 1325. Also a poor collection from Piha; Segedin 1220.

In Taylor 1263 the basidiospores were slightly larger especially in breadth, 12.5-15(-16) x 8-10(-11) µm, and may represent an independent but closely related taxon but without field data it is impossible to decide, especially as only two collections are available and each may represent the two extremes of a range of variation. In Segedin 1220 the basidiospores had a slightly excentric germ-pore.

In the literature 'Naucoria' pediades is recorded in Colenso (1886) and Massee (1898) but this taxon has undoubtedly been confused in the past (see pg 25). Colenso b 113 is a mixed collection of which the smaller element refers to A. pediades; the other part of b 113 refers to A. semiorbicularis q.v.

7. A. praecox (Pers.: Fries) Fayod in Ann. Sci. Nat. (Bot.) Series 7, 9: footnote to 358, 1889. Figs. 8 & 9.
Basionym: Agaricus (Pratella) praecox Persoon: Fries, Systema Mycologicum I: 202, 1821.
Synonym: Pholiota praecox (Pers.: Fries) Kummer, Der Führer in die Pilzkunde: 85, 1871.

Description of New Zealand collections:

Pileus (20-)33-80 mm, convex with inrolled margin, purplish
brown when very young soon yellow ochre to yellowish, browning
with age, smooth, texture of kid-glove; margin smooth with
narrow greyish border where flesh is thin, when immature
persistent purplish brown, often with velar remnants. _Stipe_
60-80 x 6-8 mm, equal or tapered upwards from swollen or
bulbous base, annulate, whitish and striate at apex, yellowish
below ring but browning on handling, solid then hollow,
attached to conspicuous white mycelial cords at base; _ring_
membranous, superior, whitish, darkening with age. _Gills_ adnate-
sinuate, crowded, pale cream-colour at first becoming pinkish
brown-grey finally darkening. _Flesh_ relatively thick, pale
cream-colour with a grey line above gills, creamy white,
fibrous in stipe. _Smell_ sour.

Basidiospores deep brown in mass, 8-9.5(-10) x (4.5-)5-6 µm,
ellipsoid, smooth, thick-walled, relatively strongly coloured
in water and aqueous alkali solutions; germ-pore central,
distinct. _Basidia_ 4-spored, 20-25 x 6-7.5 µm. _Cheilocystidia_
a mixture of clavate to vesiculose cells (30-45 x 18-20 µm) with
a few ventricose cells 15-20 x 5-10 µm; _pleurocystidia_ frequent,
prominent, mucronate, lageniform or even ampulliform, 45-55
x 18-25 µm, sometimes with minute granulations at apex.
Hymenophoral trama parallel with a few oleiferous hyphae.
Pileipellis a palisadoderm of spheropedunculate cells 17-38 µm
diam. _Clamp-connections_ present.

Material examined: Caespitose in _Asparagus_ bed, Oratia,
legit J.W. Endt, 27 ix 1965; _PDD_ 24770.   In soil with sawdust,
garden cleared from manuka bush, south branch of Waianakarua
River, 16 x 1966, legit A. Douglas, _Taylor_ 317 & 318; in grass,
Gudex Memorial Park, near Cambridge, Waikato District, 20 ix
1969, legit J.L. Austwick, _Austwick_ 1879; in grass, Soil
Research Station, Rukuhia, near Hamilton, 27 ix 1969, legit
E.P. White, _Austwick_ 1942; gregarious on ground, New Lynn,
Auckland, 30 ix 1970, legit J.S. Cole, _PDD_ 28569; on ground,
New Lynn, Auckland, x 1970, legit J.S. Cole, _Segedin_ 751;
on ground under _Quercus_, Auckland University grounds, 3 x 1971,

B.S. Parris, <u>PDD</u> 29685; on ground under <u>Magnolia</u>, Landscape Road,
Auckland, 21 ix 1980, <u>Segedin</u> 1719; in coastal paddocks, with
<u>Lupinus arboreus</u> Sims, grazed by cattle, vii 1982, legit L.R.B.
Mann, <u>Taylor</u> 1254A*; in bark covered garden, Old Biology Building,
University of Auckland, viii 1982 and ix 1982, <u>Taylor</u> 1258 (A-C);
on soil in picnic area, Huka Falls, 27 x 1982, <u>Taylor</u> 1261.

<u>Colenso</u> <u>b</u> 283 cannot be added to this list meantime although
Horak (1971b) states 'all found characters correspond well with
those of the type' i.e., <u>A. praecox</u>; see pg. 54.

A widespread agaric of gardens and fields in western Europe
from late spring until summer even into late August in more
northerly regions; probably introduced into New Zealand.
Material has been compared with that from Europe (Scotland:
<u>Orton</u> 2101-3 incl & <u>Wat.</u> 11745 in E). This is a world wide
species of cultivated and disturbed areas: it is a rather
variable species, or more likely a complex of taxa; see Watling,
1985. Cultural studies outlined on pg. 19 would be valuable.
This species has been recorded by Colenso (1886) and figured
by Massee (1898) but see pg. 54. The description is taken
from Taylor's collections although the material in <u>PDD</u> accompanied
by short descriptions agree in all major details.

A collection in <u>PDD</u> (on sawdust, Mt. Albert, Auckland City,
ix 1969, legit K.R.W. Hammett, <u>PDD</u> 29208) thought to be <u>A.praecox</u>
has processes on the pleurocystidia similar to, although less
distinct, than those in <u>A. arvalis</u>. This type of pleurocystidia
is also found in <u>A.</u> <u>acericola</u> (Peck) Singer s. Overholts and
the Hammett collection may represent this widespread N. American
taxon. See above.

A collection from Otago (amongst <u>Ammophila</u> in dunes, near
Brighton, 28 xii 1969, <u>Austwick</u> 1921) may also represent a
slender <u>A. praecox</u> but in the absence of field-data, other

---

* This collection was eaten and reported to induce hallucinations.
  This is the first time such an observation has been made with
  <u>Agrocybe</u> spp. but further investigation is required as the
  collection is known to be intermixed with at least one specimen
  (separated as 1254B) of another bolbitiaceous fungus with spores
  13.5-15.5 x 7-8 µm (Fig. 13K). The toxicity of <u>Conocybe</u> spp.
  has been recorded.

than an uninformative pencil sketch, this cannot be confirmed;
in such areas in Britain A. sphaleromorpha (Bull.: Fr.) Fayod
might be expected. Vesiculose cheilocystidia could not be
located although typical ampulliform pleurocystidia clearly
approached the gill-edge; some basidiospores in any one mount
were also rather large, e.g. 11-12 x 7-8 µm. Equally,
filamentous cheilocystidia taken as characteristic of A.cf.
puiggarii were not found; see below.

8. A.cf.puiggarii (Spegazzini) Singer in Lilloa 23: 212 (1950).
Fig. 1 A-J & O-S.
Basionym: Pholiota puiggarii Speg. in Bot. Acad. Nac. Cienc.
Cordoba 11: 413 (1889).

This species was originally described, as Pholiota based on
a collection from grassy fields at Apiahy, Brazil. Singer (1950)
recognised it as an Agrocybe and placed it close to A. praecox
and A. molesta (Lasch) Singer (= A. dura (Bolt.: Fr.) Singer).

There is evidence that this species may be more widespread
than the literature suggests and that it may occur in New
Zealand. Three collections with characteristic flexuous filamentous to lageniform cheilocystidia and mucronate-vesiculose
pleurocystidia were located amongst the material in PDD.
Unfortunately little field data accompanies the collections
so the identifications are meantime preliminary. It, however,
indicates that yet another member of Agrocybe sg. Agrocybe is
to be found in New Zealand and fresh collections to substantiate
the record are urgently required. The New Zealand material had
the following characters:

Pileus 55-90 mm, convex then plano-convex, ochraceous
yellow, paler at enrolled margin and sometimes ornamented with
velar debris, cracking from edge to form areolae and show white
flesh below, irregularly furrowed. Stipe 40-50 x 5-13 mm,
cylindric, slightly clavate towards base, white at apex,
annulate, tough, fibrillose-floccose and similarly coloured to
pileus downwards, becoming cracked; ring thick, tough,
floccose-flaky below, dark brown striate above from spores.
Gills crowded, fairly thin, dirty brown grey, uneven with pale,

wavy edge. <u>Flesh</u> white throughout almost continuous in pileus
and stipe. <u>Taste</u> none; <u>smell</u> rather sweet-sickly.

<u>Basidiospores</u> tobacco brown in mass, (10.25-)11-12.5 x
6-6.5 μm, elliptic in side and face views, slightly flattened
in side view, relatively thick-walled, yellowish brown in water
and alkali solution; germ-pore distinct, central. <u>Basidia</u>
4-spored, cylindric 27-35 x 6-9 μm. <u>Cheilocystidia</u> numerous
forming a thick wide fringe, narrowly lageniform, 40-60 x
5-10 μm with flexuous neck 25-30 μm long and obtuse head,
tapered or not and 4-6 μm broad, some almost cylindric, hyaline;
<u>pleurocystidia</u> ventricose-mucronate or ampulliform 32-46 x
10-14 μm with mucro 1-2 μm broad, hyaline. <u>Hymenophoral trama</u>
regular with some oleiferous-like, yellow hyphae. <u>Pileipellis</u>
a mixture of pyriform, pedicellate swollen cells >25 μm broad
intermixed with many filamentous-cylindric, hyaline hyphae
and fewer, similar yellowish hyphae 4-6 μm broad. <u>Stipitipellis</u>
above ring with lines of clusters of variable cells many
similar to those on gill margin or flexuous, filamentous, some
pyriform. <u>Clamp-connections</u> present.

<u>Habitat</u>: on ground or on old wood in disturbed areas.

<u>Material examined</u>: Old Government House, Auckland University
Campus, legit J. Cole, 1 iv 1971, <u>Segedin</u> 760; on wood of
<u>Senecio</u>, Old Government House, Auckland, iii 1972 <u>Segedin</u> 849;
on ground, Cornwall Park, Auckland, legit C.M. Segedin, 1 xi
1976 <u>Segedin</u> 1334; Clive Road, Epsom, 3 xii 1980, <u>Segedin</u> 1723.
Possibly also <u>Segedin</u> 690, Auckland Domain 1970?

9. <u>A. semiorbicularis</u> (Bulliard: St Amans) Fayod in Ann. Sci.
Nat. (Bot.) Series 7, 9: 358, 1889. Fig. 3 F-I.
<u>Basionym</u>: <u>Agaricus semiorbicularis</u> Bull.: St Amans, Flore
Agenaise: 56 (1821).
<u>Synonym</u>: <u>Naucoria semiorbicularis</u> (Bull.: St Amans) Gillet
Champ. France: 439 (1874).

No descriptions of the macroscopic characters are available
for New Zealand collections and therefore the determination is
based solely on microscopic characters. For the characters of
basidiomata in the field readers are referred to European

literature, e.g. Watling (1982). A. semiorbicularis is delimited
by the following combination of microscopic characters:

a. Broadly ellipsoid to ovoid basidiospores 10.5-14 x 7-8(-10) µm,

b. Basidia mostly 2-spored,

c. Absence of pleurocystidia and

d. Lageniform cheilocystidia with ± subcapitate head.

It is widespread on lawns and in grasslands.

Details of the New Zealand collection are as follows:-

Basidiospores 14-16 x (8-)8.5-11 µm, ellipsoid-ovoid, medium
ochraceous in water, darker in aqueous alkali, smooth; germ-pore
distinct. Basidia 2-spored, a few 3- and 4-spored basidia seen,
25-30 x 7.5-10 µm. Cheilocystidia 20-30 x 5-7.5 µm, ampulliform
with slightly swollen obtuse apex (3-5 µm broad); pleurocystidia
absent. Caulocystidia subcapitate to slightly swollen at apex,
20-30 x 6-8 µm, apex 4-5 µm.

Material examined: as Agaricus pediades, Napier, Colenso
b 113 pro parte & b 269; as Agaricus semiorbicularis, Napier,
Colenso 1053.

Agaricus semiorbicularis is recorded in the literature by
both Colenso (1890) and Massee (1898).

10. A. temulenta (Fries) Singer in Beih. Bot. Centralblatt, Abt.
B., 56: 167, 1936 (non sensu Singer, 1936).
Basionym: Agaricus (Galera) temulentus Fries, Systema
Mycologicum: I: 268, 1821.
Synonym: Naucoria temulenta (Fries) Kummer, Der Führer in die
Pilzkunde 22: 77, 1871.

This species is recorded in both Colenso (1890), and Massee
(1898) but see pg 55.

A. temulenta is sporadically recorded in Europe in grassy
places, often in previously disturbed areas and one of us (RW)
has seen a recent collection from North America (New York State).
It is easily recognised when young by the uniformly bright
ochraceous yellow pileus and stipe which fade and become flushed
with saffron, and the stuffed stipe with tough cortex. The
basidiospores, (10.5-)11-13(-14.5) x (6.5-)7.5-8.5(-9) µm, are
ellipsoid and basidia 2-or 4-spored.

11. <u>A. vervacti</u> (Fries) Singer in Beih. Bot. Centralblatt Abt.
B., 56: 167, 1936.
Basionym: <u>Agaricus vervacti</u> Fries in Systema Mycologicum I:
263, 1821.
Synonym: <u>Naucoria vervacti</u> (Fries) Kummer, Der Führer in die
Pilzkunde 22: 77, 1871.

   This species is recorded in Colenso (1886) but see pg 53.

   <u>A. vervacti</u> is not uncommon in Europe and might be expected
in grassland in New Zealand and if so can be recognised by the
small basidiospores, ((6-)7.5-8.5(-9) x 4.5-6 µm) with a very
small germ-pore, and the variable cheilocystidia often with a
narrow neck and head. In stature and colour <u>A. vervacti</u> resembles
<u>Stropharia coronilla</u> (Bull.: Fr.) Quél. but can be easily
distinguished by the hymeniform pileipellis, lack of chrysocystidia
and snuff brown not distinctly purple brown spore-print. See pg 7.

12. <u>Agrocybe</u> sp. 1
   A collection in K labelled <u>Agaricus erebius</u> (H1257/80) is
undoubtedly a member of sg. <u>Agrocybe</u> possessing spores (7-)
7.5-8.5(-10) x 5.5-6(-6.5) µm with a distinct truncate germ-pore.
In the <u>A. erebia</u> group the spores are fusiform to boletoid and
lack a germ-pore (i.e. sg. <u>Aporus</u> sect. <u>Velatae</u>). Horak (1971b)
also infers that this is a member of the genus <u>Agrocybe</u>. The
dark coloured pileus suggested from the original naming indicated
that this is very different from anything we know. It is too
young to extract any further information.
   The K material of seven basidiomata and one isolated pileus
possesses the following characters (compounded from all
specimens):
   <u>Basidiospores</u> (7-)7.5-10 x 5.5-6(-6.5) µm, ovoid, broadly
elliptic in face-view, slightly flattened in side-view (Fig. 7L).
<u>Basidia</u> 4-spored, c. 25 x 7.5 µm, clavate. <u>Cheilocystidia</u>
gill-margin grazed; <u>pleurocystidia</u> ventricose to shortly
lageniform, 25-35 x 9-15 m, apex 7.5-8 m broad. <u>Pileipellis</u>
damaged, a mixture of filamentous and some vesicular cells.
   Another specimen in <u>PDD</u>, 682 also labelled <u>Pholiota</u> near

erebia (on ground, Weraroa, 3 x 1919, G H Cunningham), is heavily
moulded but might represent the same species; both lageniform
cheilo- and pleurocystidia were located. It has similarly shaped
spores of parallel size and although the basidiomata are poorly
preserved similar pleurocystidia to those on the K material were
found (27-38 x 12.5-14 µm, apex 5.5-8.5 µm); cheilocystidia are
present, vesiculose to mucronate with obtuse apex, 20-25 x 13-14
µm, apex 4 µm.

13. Agrocybe sp. 2

    A collection also in K simply labelled 'Agaricus strophosus
Fries, New Zealand' is not a species of Hebeloma as would be
assumed from the use of Fries' name but a species of Agrocybe.
We are grateful to Egon Horak for drawing this Colenso collection,
which appears to be from Wairarapa (Massee, 1898), to our
attention. The collection consists of a partially expanded
basidioma with distinct velar remains on both the pileus-margin
and stipe. The gill-margin is badly damaged but the basidiospores
are characteristically bolbitiaceous, with thick-wall and large,
truncate germ-pore; the basidia are 4-spored.

    It can only be assumed that the basidioma when fresh was
relatively pale-coloured from the epithet chosen but it is
impossible without field data to trace the relationships of this
collection. The following additional data are offered:

    Basidiospores 10-12 x 7-8.5 m, ellipsoid, slightly
flattened in side-view, smooth, thick-walled, ochraceous brown
in water, darker in aqueous alkali solutions; germ-pore distinct,
broad (Fig. 7 M). Basidia 4-spored (Fig. 7 N). Cystidia not
found.

BOLBITIUS

    Basidiomata soon collapsing, delicate, coprinoid, minute to
medium. Pileus usually with brightly coloured vacuolar
pigmentation, violaceous, pinkish or yellow, rarely brown,
rarely white, 'deliquescent', campanulate, cylindrical or ellipsoid
becoming convex, finally fully expanded, viscid, smooth or wrinkled.
Stipe central, usually white or pale-coloured, rapidly collapsing

after expansion of the pileus before any darkening can take place
veil absent. <u>Gills</u> free or if adnexed at first then rapidly free
ventricose, cream, pallid to ochraceous, finally rust-colour with
distinct whitish, flocculose margin. <u>Flesh</u> thin, white or pallid.

<u>Basidiospores</u> smooth, relatively thick-walled, with distinct,
often very prominent germ-pore, rust-colour in mass. <u>Basidia</u>
4-spored, very rarely 2-spored, and in one New Zealand species
more than 4-spored, usually separated by distinct, inflated
brachycystidia. <u>Gill-trama</u> regular becoming irregular to
alveolate with age. <u>Cheilocystidia</u> distinct, vesiculose to
inflated, sometimes with differentiated neck; <u>pleurocystidia</u> if
present usually similar to cheilocystidia. <u>Pileipellis</u> a distinct
palisadoderm covered with a gelatinous pellicle; 'scalp'
cellular. <u>Stipitipellis</u> of hyaline, cylindric, parallel
cells covered in part or throughout in caulocystidia similar
to those on gill-margin.

Development paravelangiocarpic.

On the ground in woods, pastures, heaths etc., on dung, plant
remains and refuse, and on wood; saprophytes. Those species on
wood have often been placed in the separate genus <u>Pluteolus</u>.

One species <u>B. muscicola</u> appears to be endemic and widespread
in mixed forest on very soft wood in which it produces a stringy
white rot; the other common species is <u>B. vitellinus</u>. Two
further collections are at the moment unplaced.

Key to Species

1. Pileipellis composed of rounded and <u>Opuntia</u>-
   like cells; pileus sepia or purplish brown
   with depressed patches (scrobiculate)   1. <u>B. muscicola</u>
1x Pileipellis composed of rounded and
   spheropedunculate cells only; pileus yellow,
   or whitish with traces of yellow at disc   2
2. Stem whitish, lemon-yellow or at most pale-
   coloured   3
2x Stem distinctly egg-yellow from beginning   <u>Bolbitius</u> sp. 1
3. Pileus white, yellowish or buff only at
   disc with age   3. <u>Bolbitius</u> sp. 2
3x Pileus more strongly coloured   4

4. Pileus delicate, membranaceous, lemon chrome to
   lemon yellow, soon clay buff to vinaceous buff;
   gills cream then cinnamon buff; basidiospores
   often >7 μm broad   2. B. titubans
4x Pileus usually thick at disc, egg-yellow or
   chrome yellow; gills straw then red-brown;
   basidiospores generally <7.5 μm broad   3. B. vitellinus
   (if with septate caulocystidia and numerous
   vesiculose cheilocystidia see Bolbitius sp.1
   and   if with 5-spored basidia see Bolbitius
   sp. 2)

1. B. muscicola (Stevenson) Watling in Bibl. Mycol., 82: 80,
1981. Figs. 10 & 11.
Basionym: Pluteus muscicola Stevenson in Kew Bull. 16: 72, 1962
Synonym: Pluteolus muscicola (Stevenson) Horak in New Zealand
Journ. Bot. 9: 438, 1971.

   Pileus 18-50 mm, membranaceous, convex soon plano-convex or
plane, sometimes umbonate at first, then depressed, very thin and
delicate, thick-gelatinous then glutinous from clear or faintly
sienna, watery jelly, whitish, pale buff, vinaceous buff or smoke
grey beneath gluten with radially arranged, dark brown or black
depressions, smaller and less conspicuous towards substriate
margin, and paler areas between forming an anastomosing network;
in some specimens the resulting pattern resembling a leopard
skin. Stipe 17-50 x 2-3 mm, cylindric or slightly swollen towards
base, narrowest at centre, brittle, white, pruinose throughout or
only at apex, and then appressed fibrillose below, base sometimes
tomentose, hollow. Gills almost free or adnexed, pale orange
brownish then with rust tinge, papery, very thin and fragile,
fairly crowded with fimbriate, paler edges. Flesh extremely
thin in pileus, whitish or watery, very translucent near pileus-
margin, silky fibrous in stipe, yellowish towards stipe-base.
Smell none.

   Basidiospores orange-tawny in mass, 7.5-8.5(-9.5) x 4.5-5.5(-6)
μm ellipsoid, very slightly amygdaliform in side-view, smooth,
relatively thick-walled, truncate from broad germ-pore. Basidia
4-spored, clavate-pedicellate, hyaline, in 'pavement' with shorter

brachycystidia. <u>Cheilocystidia</u> lageniform with short to
elongate neck, 17.5-35 x 6-10.5(-12) μm, apex slightly swollen
2.5-5(-7) μm; <u>pleurocystidia</u> absent. <u>Pileipellis</u> a mixture of
chains of inflated cells, 14-22.5 μm long, and vesiculose to
spheropedunculate cells, 29-40 x 15-26 μm, some end-cells filled
with pale greyish to lilaceous brown vacuolar sap, long, flexuous,
septate hyphae also present pushing through gelatinous pellicle.

<u>Material examined</u>: on wood, rooting amongst mosses, Tararua
Range, Tararua, Levin, 18 vi 1949, <u>Stevenson</u> 656 (holotype in K);
on rotting log of <u>Nothofagus</u>, Lowry Bay, Wellington, 23 iv 1961,
<u>Taylor</u> 84; on rotting log in mixed broadleaved forest (<u>Coprosma</u>,
<u>Griselinia</u>, <u>Fuchsia</u>) Akatore, Dunedin, 6 xii 1966, <u>Taylor</u> 319;
on very rotten wood, mixed forest, Trotters Gorge, 25 iv & 23 x
1967, <u>Taylor</u> 342 & 342B; on wood in <u>Nothofagus</u> and broad-leaved
forest, Woodside, 29 iv 1967, <u>Taylor</u> 347; on <u>Nothofagus fusca</u>
Oerst., Eglinton, 10 xi 1967, <u>Horak</u> ZT 67/192; Kowai River, Mt.
Grey, 31 xii 1968, <u>Horak</u> ZT 68/690; on <u>Nothofagus cliffortioides</u>
Oerst., Craigieburn, 5 ii 1969, <u>Horak</u> ZT 69/46; on very rotten
log, probably <u>Dacrydium</u>, rimu-rata-kamahi forest, Lake Wilkie,
Tautuku, 24 iv 1971, <u>Taylor</u> 665; on <u>Nothofagus</u> log, Brian Duder's
bush, Clevedon, 16 iv 1983, <u>Taylor</u> 1322.

The spotting, pattern of raised lines and scrobiculae of the
pileus and their colouring is very variable, and probably depends
on age of the basidioma and environmental conditions, although
the microscopic characters are relatively constant. However,
a collection on <u>Nothofagus fusca</u> (<u>Horak</u> ZT 67/192) differed in
the cheilocystidia being slightly forked at their apex, the
basidiospores slightly paler in overall pigmentation and slightly
smaller, and the pileipellis apparently lacking numerous elongate
hyphae. The collection, however, comes well within the variation
accepted for <u>B. muscicola</u>.

Horak (1971b) agrees with our disposition of Stevenson's taxon
although he prefers to use the generic name <u>Pluteolus</u>. Stevenson's
material differs only in that a few 2-spored basidia were found;
this undoubtedly explains the slightly larger spore-size for the
type material; also see Stevenson, 1982a. ZT 69/46 is somewhat
different in that although the immature pilei are scrobiculate

the old pilei have dark raised lines upon them.

2. B. titubans (Persoon: Fries) Fries, Epicrisis Systematis
Mycologici: 254 (1838). Fig. 12 F & G.; Fig. 13 A-J.
Basionym: Agaricus titubans Bull.: Fr., Systema Mycologicum 1:
304 (1821).

Several collections examined possess the delicate facies of an
agaric which in Europe has been interpreted as B. titubans (Watling,
1982). Without more extensive field data, however, the majority
of records cannot be substantiated. A collection from Piha Valley
(1 iv 1974, legit J.C. Segedin, Segedin 1171) and one from above
Huka Falls (on mown grass-straw, beside Waikato River, 27 x 1982,
Taylor 1260) possessed the broader basidiospores associated with
B. titubans. Taylor 190 (with grass, roadside bank, Holmes St.,
Oamaru, 26 iv 1964, legit L R Taylor) is immature but may represent
the same taxon. In the Huka Falls collection some development
of biporate spores was observed, something which was found to
an even greater degree in Segedin 1161 (in grass outside entrance
to Kauri Glen, Auckland, 28 iv 1974). Unfortunately the little
field data available only indicates that the agaric is coloured
much the same as members of the B. vitellinus group. The
microscopic data are as follows:-

Basidiospores 10.5-13(-14) x 6-7.5(-8.5) µm, ellipsoid to
heartshaped, lobed, lenticular, smooth, thick-walled, strongly
pigmented (very dark brown); germ-pore prominent, broad,
frequently paired. Basidia 1-,2-,3-, 4-spored, mostly 2-spored,
20-25 x 9-10 µm, Cheilocystidia (30-)35-60 x 9.5-18 µm, clavate-
vesiculose, spheropedunculate, hyaline, some twinned - (rabbit-ear
shaped); pleurocystidia absent. Pileipellis of clavate to
spheropedunculate hyaline cells, often with long pedicel, 35-50
x 10-15 µm. Clamp-connections absent.

The large number of bi-porate basidiospores resemble those of
many members of the strophariaceous gastromycetoid series;
similar spores have been found in collections of Agrocybe from
Kashmir and Poona, India (Watling & Abraham, in press). This
fungus needs to be refound in New Zealand. In some ways the
collection resembles Gastrocybe lateritia Watl. and G. incarnata

(Peck) Baroni, although it is a typical agaric. This is yet
another example of a bolbitiaceous fungus which reduces the
hiatus between the true agarics and the so-called gastromycetes
(secotioid). The Segedin collection lacks the colour typical of
G. lateritia and G. incarnata but resembles the fungus named
(ad interim) as Bolbitius rogersii (Heim, 1968); see Watling,
Quadraccia & Tabarés (ined).

The reader is referred to the discussion under B. vitellinus
of which B. titubans has variously been considered a synonym,
variety or subspecies (Watling & Gregory, 1981); Bell's record
(1983) of B. vitellinus may refer here.

3. B. vitellinus (Persoon: Fries) Fries, Epicrisis Systematis
Mycologici: 254, 1838. Fig. 12 A-D & H.
Basionym: Agaricus vitellinus Pers.: Fries, Systema Mycologicum
I; 303, 1821.

Description of New Zealand collections:

Pileus conical (15-30 mm), rapidly becoming plano-convex or
plane (15-60 mm), with or without an umbo, bright yellow (yellow-
chrome) darker at the disc, paler towards the margin where it
becomes cinnamon-brown, shining, glutinous, finally fawn throughout except for yellow centre; margin striate, splitting and
partially collapsing. Stipe 70-110 x 2-3 mm, tapering upwards,
sometimes somewhat wider at the base, whitish or extremely pale
yellow, pruinose or farinaceous throughout becoming white-silky,
stuffed. Gills free or adnexed, pale yellowish with paler watery
margin, becoming fawn, finally red-brown, fairly crowded, very
thin and papery. Flesh very thin, distinction between pileus
and stipe obvious, whitish or pale yellowish.

Basidiospores 11.5-14 x 7-8.5 μm, ellipsoid, thick-walled,
sienna in water more strongly pigmented in aqueous alkali solutions,
smooth; germ-pore large, central. Basidia 4-spored, a few
2-spored, clavate-pedicellate, 21-24 x 9.5-13 μm, hyaline, intermixed with sterile brachycystidia (pavement cells) 14-17.5 μm
broad. Cheilocystidia irregularly lageniform with short or
long and then flexuous neck, some simply vesiculose, 23-36(-48)
x 9.5-12(-19) μm, hyaline; pleurocystidia infrequent, utriform

or shortly lageniform, 23.5-28.5 x 7-10 µm. <u>Pileipellis</u> of spheropedunculate cells. <u>Stipitipellis</u> of hyaline, cylindric hyphae with clusters of irregular clavate, lageniform to vesiculose cells, especially at apex.

<u>Material examined</u>: amongst grass, Waikanae, N of Wellington, 2 viii 1958, legit M. Bulmer, <u>Stevenson</u> 1414 (notes in K); in sawdust enriched garden soil, recently cleared manuka bush, near McKerrows farm, south branch Waianakarua River, North Otago, 10 x 1966, legit A. Douglas, <u>Taylor</u> 313; under <u>Salix</u>, <u>Cordyline</u> and mixed shrubs, in limestone, beside Oamaru Creek, Devils Bridge, Oamaru, 12 iii 1970, legit L.R. Taylor, <u>Taylor</u> 570.

<u>B. vitellinus</u> is widespread in temperate Europe and N America and along with <u>B. titubans</u> could well have been introduced into New Zealand with stock, feed-stuffs or allied material; see Stevenson, 1982a and Bell, 1983. It grows naturally on a whole range of substrates and so could easily survive in a new area. The increased frequency of biporate basidiospores in collections of <u>B.titubans</u> and the appearance of 5-spored basidia in <u>Taylor</u> 246 (see below) may indicate some genetic inbalance in certain New Zealand collections.

<u>B. vitellinus</u> is a very variable fungus for which several varieties and subspecies have been recognised depending on size and growth pattern, although in many cases these are probably simply an expression of the food status of the substrate. After further field work <u>Bolbitius</u> sp. 1 (ZT 69/109), described below, may be found to be included within the concept of this taxon. At present the persistently yellow stipe is significant.

4. <u>Bolbitius</u> sp. 1.

Material collected by Horak (ZT 69/109) from Tophouse Saddle, Nelson may represent a new taxon within the <u>B. vitellinus</u> group or an extreme form.

<u>Pileus</u> conical becoming convex, umbonate-expanded or plane, yellow (egg-yolk colour), wrinkled when young, conspicuously striate-sulcate splitting at margin, fragile, membranaceous. <u>Stipe</u> cylindrical concolorous with pileus or yellowish, dry, whitish pruinose at apex, hollow, fragile. <u>Gills</u> free (to

adnexed), ventricose, intermixed with lamellules, yellowish
then turning rust, with even, whitish or concolorous edge. Taste
and Smell none.

Basidiospores 10-12.5 x 6-7.5 µm, elliptic-ovoid, slightly
flattened in side-view, red-brown in alkali solutions, thick-walled;
germ-pore central. Basidia 4-spored, hyaline, clavate, pedicellate,
25-28 x 9-11 µm, (including sterigmata ⩽3 µm long). Cheilocystidia,
numerous, polymorphic varying from vesiculose (25-30 µm broad) to
utriform (12-15 µm broad), to clavate or catenulate, 30-50 µm
long; pleurocystidia absent. Pileipellis a hymeniderm of
spheropedunculate cells 40-50 µm broad. Stipitipellis of
cylindric hyphae supporting clusters of vesiculose, hyaline
caulocystidia intermixed with chains of cells similar to gill-
edge, 16-20 x 35-65 µm.

The persistently yellow stipe is significant, and microscopically
this collection can be characterised by the innumerable, inflated
vesiculose, hyaline cheilocystidia and polymorphic frequently
septate caulocystidia. B. vitellinus generally has a cream,
straw or yellowish buff stipe; B. variecolor Atk. has a deeper
yellow stipe and the pileus is olivaceous.

5. Bolbitius sp. 2. Fig 12E.

Pileus 25 mm, conical to plano-convex with a slight central
umbo, white with yellowish disc, very thin especially towards
margin, radially grooved near margin and finally split to show
gills. Stipe 7 mm long, fawn with dull white, waxy tomentum at
base, very fragile. Gills adnexed, yellowish becoming tawny
brown. Flesh thin, whitish.

Basidiospores 13-15.5 x 7-9.5 µm, ellipsoid, thick-walled,
smooth, strongly pigmented; germ-pore distinct. Basidia 4- to
5-spored, clavate-pedicellate, 22-24 x 11-12 µm, hyaline.
Cheilocystidia utriform to ventricose or simply vesiculose,
18.5-36 x 10.5-15.5 µm; pleurocystidia absent, one broadly
lageniform cystidium seen near to gill-margin.

Material examined: solitary, on soil under Coprosma, kanuka
broad-leaved forest. Trotter's Gorge, North Otago, 15 i 1966,
legit M. Taylor, Taylor 246.

This is a most unusual agaric as several 5-spored basidia were
located in the hymenium. One of us (RW) has examined literally
thousands of basidiomata of members of the Bolbitiaceae without
observing this phenomenon before. It would be exciting to
refind this fungus and see if this character is constant.

CONOCYBE

Basidiomata usually slender, delicate, mycenoid, never
deliquescent, small to medium (5-25 mm) only occasionally in
non-New Zealand taxa larger, uniformly brown, less frequently
white although some extralimital taxa may be coloured, but never
with vacuolar sap. Pileus campanulate, cylindrical or ellipsoid
to convex, rarely expanding, hygrophanous or expallent, greasy
or moist, rarely viscid, often glistening when dry or becoming
wrinkled, smooth or pubescent, usually striate when moist,
slightly sulcate or not, more rarely plicate-sulcate, some
species with appendiculate veil. Stipe central, either striate
or silky, often villose, pubescent or pruinose in most species,
white or some shade of brown especially with age; some species
annulate or with silky fibrillose floccules. Gills adnate, only
occasionally free or with tooth, at first strongly ascending,
linear, lanceolate. Flesh usually thin, concolorous or paler
than pileus and stipe, although often darkening in stipe from
base up with age.

Basidiospores smooth, thick-walled with distinct or indistinct
germ-pore. Basidia 4-spored. Hymenophoral trama regular
becoming less regular with age, of inflated cells surrounding
a thin almost obliterated, central, filamentous strand.
Pileipellis a distinct palisadoderm with or without distinct
dermatocystidia and/or hair-like cells (see pg 45).
Stipitipellis of parallel, cylindric hyphae with brown cell-walls;
caulocystidia often numerous resembling those on gill-margin, or
slightly modified and less distinctly differentiated, accompanied
or not with hair-like cells.

Development where known in European species paravelangiocarpic
or modifications of this.

On the ground in woods, pastures, gardens etc. and on dung;

only occasionally directly attached to plant-remains or vegetable
refuse; saprophytes.

Key to Subgenera

1. Basidioma with distinct remnants of veil   2

1x Basidioma lacking ring, volva or appendiculate
    veil remnants at margin of pileus   3

2. Basidioma not volvate but with distinct velar
    remnants on stipe or at pileus-margin   *Pholiotina*
    (*Descolea* (Cortinariaceae) would key out
    here if spore ornamented see pg. 55)

2x Basidioma with a distinct volvate base   *Conocybe* sect.
    (Not yet known from New Zealand)   *Singerella*

3. Cheilocystidia lecythiform   4

3x Cheilocystidia various but if capitate
    <u>not</u> lecythiform   5

4. Basidiospores roughened   *Ochromarasmius*
    (In some parts of Asia *Conocybe* spp. with
    nodulose basidiospores are found:
    sg. *Conocybe* sect. *Nodulososporae*)

4x Basidiospores smooth   *Conocybe*

5. Pileus plicate-striate; cheilocystidia
    with long flexuous neck   *Galerella*
    (Not yet recorded from New Zealand)

5x Pileus matt or rugulose, or if striate
    then not plicate; cystidia lacking
    long neck   *Piliferae*

The only subgenus not as yet found in New Zealand is *Galerella*
but it should be looked for there in grassland, on lawns, and
close to human habitation. Subgenus *Conocybe* sect. *Singerella*
has been recorded from New Guinea and Malaysia; and sect.
Nodulososporae from Japan and Malaysia.

The genus is probably widespread but easily overlooked as
'little brown agarics'. The species are defined not only by the
fresh colours of the pileus and stipe but also by the shape and
distribution of the cheilo- and caulocystidia. Fourteen distinct
entities have been recognised, of which ten have been named. Nine
are placed in sg. Conocybe; three are new.

Key to Species

1. Stipe with distinct ring; basidiospores smooth 2
1x Stipe lacking velar remnants, caulocystidia
   often prominent throughout; basidiospores
   smooth or minutely roughened 4

2. Pileus <12 mm, ochre brown;   basidiospores 8.5-10
   x 5-6 µm;   stipe minutely pubescent at apex,
   almost smooth below ring;   basidioma
   slender 1. C. gracilenta
2x Pileus >12 mm;   spores larger, $\geqslant$10 µm, or if
   smaller then pileus dark coloured 3

3. Basidiospores 8.5-9.5(-10.5) x 5-5.5 µm 9. C. rugosa
3x Basidiospores 10-12.5 x 6-6.5 µm 10. C. aff. vexans
   (If spores 11-13 x 5.5-6 µm, slightly
   fusiform in face-view and pileus
   strongly wrinkled at disc see notes
   under C. vexans)

4. Cheilocystidia cylindric-lageniform,
   if subcapitate then never lecythiform 5. C. novaezelandiae
4x Cheilocystidia lecythiform 5

5. Basidiospores minutely ornamented 2. C. horakii
5x Basidiospores smooth 6

6.   Pileus white or at most slightly ochraceous
     conico-campanulate, only slightly
     expanding, often collapsing and then
     resembling _Bolbitius_ q.v.   3. _C. huijsmanii_
6x   Pileus pigmented, in some shade of
     brown, always expanding to some degree   7

7.   Basidiospores lacking germ-pore   6. _C. piloselloides_
7x   Basidiospores with prominent, broad,
     central germ-pore   8

8.   Basidia 2-spored   9
8x   Basidia 4-spored (also see note on
     C. tenera pg. 51)   10

9.   Basidiospores 15.5 - 17.5 x 8.5-9.5 µm;
     caulocystidia a mixture of
     lecythiform and hair-like cells   11. _Conocybe_ sp. 1
     (If basidiospores slightly smaller
     ($\leqslant$16.5 µm) see 12. _Conocybe_ sp. 2)
9x   Basidiospores (11-)12.5-16.5(-17.5) x
     7-10 µm   see 8. _C. rickenii_

10. Basidiospores $\leqslant$10 µm long   11
10x Basidiospores $\geqslant$12.5 µm long   12

11. Basidiospores 7.5-9 x 5-5.5 µm, ellipsoid;
     caulocystidia entirely lecythiform   4. _C. mesospora_
     (If basidiospores 8-9 x 5-6 µm see
     notes with 8. _C. rickenii_)
11x Basidiospores 8-9(-10) x (5.5-)6-6.5
     x 5.5-6 µm, lenticular, hexagonal to
     angled, broadly elliptic in face-view   13. _Conocybe_ sp. 3

12. Basidiospores 15-16 x 8.5-9 µm; ellipsoid;

    caulocystidia lecythiform intermixed with

    hair-like cells   7. C. pubescens

    (If basidiospores 17-20 x 10-13 µm; see 8.

    C. rickenii and notes with 7. C. pubescens)

12x Basidiospores broadly ellipsoid or ovoid;

    caulocystidia unknown   14. Conocybe sp. 4

1. C. gracilenta Watling & Taylor, sp. nov. Fig. 14 O-R

    Pileus 10.5-11 mm, hemisphericus deinde convexus,

ochraceobrunneus, submicaceus, hygrophanus, in siccitate vel

aetate pallescens, ad discam subrugosus, ad marginem substriatus.

Stipes 33 x 1(-1.5) mm, cylindricus, ad basem subbulbosus,

pileo concolore, apice minute pruinoso, deorsum fibrillosus,

siccus, fragilis; annulo striato subpersistente. Lamellae

adnexo-adnatae, ventricosae, ochraceo-brunneae, marginibus

concoloribus vel albidis. Caro inodora et insipide.

    Basidiosporae 8.5-10 x 5-6 µm, ellipsoideae. Basidia

4-sporigera. Cystidia aciei lamellarum lageniformia, 30-40 x

8-12 µm ad apicem 2.5-3 µm. Cellulae cuticulae pilei

pyriformes vel spheropedunculatae 18-24 µm diam. Cystidia

stipitis similia.

    Ad terram. Holotypus: Horak ZT 68/303, 24 iv 1968, Kaituna

Valley, New Zealand.

    Pileus 10.5-11 mm, hemispheric to convex, ochre brown, paler

with age or when dry, strongly hygrophanous, submicaceous shiny,

disc somewhat venose, dry; margin indistinctly striate. Stipe

33 x 1(-1.5) mm, cylindric, sometimes with subbulbous base,

annulate, concolorous with pileus, apex minutely pruinose,

towards base fibrillose, dry, fragile; ring striate, subpersistent.

Gills adnexed to adnate, ventricose, pale ochre brown, with

concolorous or whitish, even edge. Taste and smell none.

    Basidiospores 8.5-10 x 5-6 µm, elliptic in face-view, slightly

flattened in side-view, thick-walled, non-amyloid, ochraceous

brown in water and aqueous alkali solutions; germ-pore prominent

and broad. Basidia 4-spored, clavate with short pedicel, hyaline

in water and alkali solutions, 22 - 25 x 8-10 µm. Cheilocystidia

lageniform with narrow venter 30-40 x 8-12 µm, neck 10-15 µm
long, not distinct, tapered to apex 2.5-3 µm, some cells almost
pointed; _pleurocystidia_ absent. _Pileipellis_ a strict palisadoderm
of spheropedunculate cells with smooth or slightly encrusted
pedicels, 18-24 µm broad. _Stipitipellis_ of parallel, brown-walled
hyphae, covered at apex with ventricose-rostrate caulocystidia,
20-35 x 4-7.5 µm. _Clamp-connections_ present.

Material examined: on soil under _Podocarpus_, _Fuchsia_ etc.,
Kaituna Valley, Banks Peninsula, 24 iv 1968, _Horak_ ZT 68/303.

This comes close to both the european C. filaris (Fr.) Kühner
and South American C. austrofilaris (Singer) Watling; it differs
from the latter in lacking the variable cheilocystidia and the
encrusted amorphous pigment on the gill-margin found in that
species. It differs also from the former particularly in its
more slender and elegant nature and the narrower cheilocystidia.
In stature it resembles C. flexipes Watling but this species has
much larger basidiospores.

2.   C. horakii Watling & Taylor, sp. nov. Fig. 2 F-N

Pileus 3.5-8.5 mm, hemisphericus vel convexus, deinde
plano-convexus, atrobrunneus, argillaceus vel melleus, in
siccitate et aetate pallescens, margine conspicue striato
interdum subsulcato. _Stipes_ 11-23 x 0.5 mm, cylindricus,
ochraceo-argillaceus deinde ferrugineus, glaber, fragilis,
fistulosus. _Lamellae_ subliberae, subventricosae, ferrugineae,
acie concolore. _Basidiosporae_ 6.5-7.5 x 3.5-4 µm, amygdaliformes,
minute punctatae, in parte superiore   attenuatis usque ad porum
angustum sed manifestum. _Basidia_ 4-sporigera. _Cystidia aciei
lamellarum_ lecythiformia 18-25 x 8-10 µm, capitula 3.5-5.5 µm.
Cellulae cuticulae pilei pyriformes vel spheropedunculatae et
lecythiformes graciles vel tibiiformes. _Cystidia stipitis_
caespitosa, tibiiformia vel lecythiformia, 15-25 x 6-8(-10) µm,
capitulo 3.5-5 µm.

Ad lignum putridum nothofagineum. _Holotypus_: _Horak_ ZT 68/66,
10 ii 1968. Westland, New Zealand.

Pileus 3.5-8.5 mm, hemispheric to convex, becoming expanded
to flat, dark brown, argillaceous or honey brown at disc, colour

fading towards margin and on drying, surface dry, glabrous; margin
conspicuously striate sometimes subsulcate with age. <u>Stipe</u> 11-23
x 0.5 mm, equal, pale ochraceous or argillaceous becoming darker
rust-brown with age, apex pruinose otherwise glabrous, appressed
fibrillose towards base, dry, hollow, brittle, with weft of white
hyphae at base. <u>Gills</u> free, subventricose, intense rust-brown,
L 8-12, l 3, with concolorous, even or subfimbriate edge. <u>Smell</u>
none.

<u>Basidiospores</u> 6.5-7.5 x 3.5-4 µm, amygdaliform in side-view,
slightly drawn out at germ-pore, minutely punctate (asperulate),
slightly thick-walled, non-amyloid, ochraceous tawny in water,
slightly darker in aqueous alkali; germ-pore distinct. <u>Basidia</u>
4-spored, clavate with short pedicel, hyaline in water and alkali
solutions. <u>Cheilocystidia</u> lecythiform, 18-25 x 8-10 µm,
capitulum 3.5-5.5 µm, hyaline; <u>pleurocystidia</u> absent. <u>Pileipellis</u>
a palisadoderm with or without dermatocystidia, composed of
generally smooth spheropedunculate cells 15-18(-20) µm broad,
sometimes with slightly roughened pedicels; <u>pilocystidia</u> when
present, lecythiform, narrow, and slightly pigmented, sometimes
even tibiiform in overall shape. <u>Stipitipellis</u> of cylindric
cells covered especially at apex with groups of tibiiform to
lecythiform caulocystidia 15.25 x 6-8(-10) µm, generally
narrower and longer than cheilocystidia. <u>Clamp-connections</u>
present.

<u>Material examined</u>: on humus with rotten woody debris of
<u>N. fusca</u>, Lake Ahaura, Kopara, Westland, 10 ii 1968, <u>Horak</u> ZT
68/66 (Holotype); on rotten <u>Nothofagus fusca</u> Oerst. wood,
Matakitaki, Murchison, Nelson, 27 i 1969, <u>Horak</u> ZT 69/24.

This species can be recognised by its very small size and
the basidiospores which are slightly ornamented (asperulate).
It resembles the European <u>C. dumetorum</u> (Vel.) Svreček in general
facies, although it is smaller, and microscopically it differs
in the structure of the basidiospore wall. Only phase contrast
microscopy accentuates the ornamentation which can be easily
missed; the basidiospores are narrower than those for
C. dumetorum. Kühner (1935) has described a form 'spores
banales' of <u>C. laricina</u> (Kühner) Kühner which might be confused

with the New Zealand fungus but the spores are broader and have a
more protruding germ-pore; C. laricina is conspecific with
C. dumetorum. Members of sg. Ochromarasmius are known from South
America (Singer, 1969), the Caribbean (Dennis, 1953) and north
temperate countries (Watling, 1976 & 1983).

3. C. huijsmanii Watling, in Nordic Journal of Botany 3: 262
(1983). Fig. 3 J-P; Fig. 16 O,N,T & V.
Basionym: Galera lactea J E Lange f. semiglobata J E Lange in
Dansk Bot. Arkiv. 9(6): 35(1938).

   Description of New Zealand collection:

   Pileus ≤20 mm (-25 mm when expanded) conical, collapsing
on expansion, white with faint central ochre flush in age, striate-
grooved in outer half. Stipe 60-70 x 1-2 mm, tapered upwards,
abruptly bulbous at the base, white satiny with traces of mealy
granules, twisted, hollow, very fragile. Gills subfree, crowded,
very thin in substance, sienna. Flesh thin and translucent in
pileus and stipe. Spore-print sienna.

   Basidiospores (12-)12.5-14 x (7-)8.5-9.5 µm, broadly
ellipsoid to almost ovoid, thick-walled, strongly pigmented in
water and aqueous alkali solutions; germ-pore prominent, large.
Basidia 4-spored, hyaline, clavate with distinct pedicel,
separated by distinct brachycystidia. Cheilocystidia lecythiform,
21-27 x (7-)9-11 µm, capitulum 3.5-4.5 µm diam.; pleurocystidia
absent. Pileipellis a palisadoderm of spheropedunculate cells
16-19 µm broad x <38 µm long.

   Material examined: in grass lawn, St Heliers, Auckland,
5 xii 1981, Taylor 1151A & B(AK1 & 2).

   This taxon is widespread in north temperate countries and
probably was introduced into Australia and New Zealand.

4. C. mesospora (Kühner ex) Kühner & Watling in Notes Roy. Bot.
Gdn. Edinb. 38: 336, 1980. Fig. 15 I-M; Fig. 16 U,W & X.

   Description of New Zealand material:

   Pileus 6-20 mm, convex to campanulate becoming expanded,
argillaceous when wet, to paler honey-brown on drying,
hygrophanous, dry, pruinose under hand-lens; margin conspicuously

striate when young. <u>Stipe</u> 22.5-42 x 1-1.5 mm, base 2.5-3 mm,
cylindric with subbulbous base, pale honey-brown, hollow, ± smooth
but pruinose above and appressed fibrillose towards base. <u>Gills</u>
adnexed, ventricose, whitish or pale becoming ochraceous rust-
colour, with whitish, fimbriate edge. <u>Flesh</u> concolorous. <u>Taste</u>
and <u>smell</u> not distinctive.

<u>Basidiospores</u> (7.5-)8.5-9(-9.5) x 4-5 µm, elliptic in face-view,
slightly flattened in side-view, relatively thick-walled, ochraceous
brown in both water and aqueous alkali solutions, darker in the
latter, non-amyloid, smooth; germ-pore prominent and broad.
<u>Basidia</u> 4-spored, hyaline, clavate with very short pedicel,
18-22.5 x 7-9 µm. <u>Cheilocystidia</u> lecythiform, 15-21.5 x 7.5-12.5
(-15) µm, capitulum 3-4.5 µm diameter, hyaline; <u>pleurocystidia</u>
absent. <u>Pileipellis</u> a palisadoderm of spheropedunculate to
ellipsoid cells, 18-40 x 7.5-12 µm with slightly darkened pedicels
and tibiiform dermatocystidia; <u>pilocystidia</u> 25-40 x 4.5-10 µm,
capitulum 3-4 µm. <u>Stipitipellis</u> of filamentous, parallel hyphae
covered in lecythiform caulocystidia, 20-25 x 10-12.5 µm,
capitulum 3-4 µm. <u>Clamp-connections</u> present.

<u>Material examined</u>: amongst debris, on rotten wood of <u>Podocarpus</u>
<u>dacrydioides</u> A. Rich., S of Ahaura, Westland, 14 iii 1968, <u>Horak</u>
ZT 68/154; on rotting leaves, under broadleaved trees, Harihari,
Westland, 13 ii 1969, <u>Horak</u> ZT 69/67.

These collections agree microscopically in all ways with
European material.

5. <u>C. novaezelandiae</u> Watling & Taylor, sp. nov. Fig. 15 B, G & H.

<u>Pileus</u> <25 mm conicus vel campanulatus, ochraceo-flavidus, ad
discum fulvus, siccus, impolitus. <u>Stipes</u> 50 - 55 mm, cylindricus
sed diameter variabilis, tortilis, fibrosus, politus, ad apicem
pallidus deorsum fuscans, farctus, facile ab pileo secedens.
<u>Lamellae</u> brunneo-ochraceae, subliberae; acie alba serrataque.
<u>Caro</u> pilei straminea tenuisque; stipitis concolora.

<u>Basidiosporae</u> ellipsoideo-amygdaliformes, 9.5-10.5 x 5.6 µm.
<u>Basidia</u> 4-sporigera. <u>Cystidia aciei lamellarum</u> irregulariter
lageniformia 29-54 x 7-10.5 µm, ad apicem 3.5-6 µm. <u>Cellulae</u>
<u>cuticulae pilei</u> pyriformes vel spheropedunculatae vel lageniformes

- 43 -

23.5-36 x 6-9.5 µm. _Cystidia_ stipitis similia.

Ad terram in pascuo ovino. _Holotypus_: _Taylor_ 178, 12 iv 1964, South Oamaru, New Zealand.

Pileus 25 mm conical or campanulate, ochraceous yellow, darker and more tawny at disc, dry, dull. _Stipe_ 50 - 55 mm, equal, but variable in thickness, flattened near apex, in one larger basidioma wider at gill-attachment, slightly twisted, shining, pale above, darkening towards the base, fibrous, pubescent throughout, cottony towards base, easily separable from pileus, stuffed. _Gills_ adnexed, brownish ochre, with white, serrate edge. _Flesh_ pale yellowish, thin in pileus, concolorous in stipe.

Basidiospores bright orange-brown in mass, 9.5-10.5 x 5-6 µm, ellipsoid-amygdaliform, pale to medium ochraceous in water, darker in aqueous alkali solutions; germ-pore present, central. _Basidia_ 4-spored, clavate, 21-29 x 7-9.5 µm. _Cheilocystidia_ irregularly lageniform, 29-54 x 7-10.5 µm, head 3.5-6 µm, sometimes with secondary swelling half-way down neck or with side-branch. _Caulocystidia_ in radiating clumps, lageniform, with less markedly elongate neck than in cheilocystidia, 24-48 x 8-12 µm. _Pileipellis_ a palisadoderm of spheropedunculate cells intermixed with a few shortly lageniform pilocystidia, 23.5-36 x 6-9.5 µm, neck 4.5-6 µm broad. _Clamp-connections_ present.

Material examined: in sheep-paddock, Maude St., South Oamaru, 12 iv 1964, legit Graham, _Taylor_ 178.

A very distinctive member of _Conocybe_ sg. _Piliferae_ related to the North American _C. capillaripes_ (Peck) Singer, and the European _C. aberrans_ (Kühner) Kühner, and closer to the latter. The former differs in its more slender stature and larger spores 12-15.5 x 7.5-10(-10.5) µm (from type-material) and the latter particularly in its more ellipsoid amygdaliform basidiospores, more campanulate pileus and habitat preferences, viz. burnt patches. The basidiospores of _C. aberrans_ are similar, viz. 8.5-10(-12) x 5-5.7(-6.5) µm.

6. <u>C</u>. aff. <u>piloselloides</u> Watling in Notes Roy. Bot. Gdn. Edinb.
40: 549, 1983. Fig. 15 A, C-F.

Description of New Zealand material:

<u>Pileus</u> <35 mm, convex-umbonate to plano-convex but retaining
umbo, yellow-brown or tawny brown, drying paler (yellowish), disc
rather darker, moist, matt, radially wrinkled about umbo: margin
radially striate, splitting. <u>Stipe</u> 35-40 x 3 mm, equal but for
abruptly bulbous base, concolorous or paler than pileus, hollow.
<u>Gills</u> adnexed, yellow-brown, crowded. <u>Flesh</u> extremely thin in
pileus, whitish, distinct from the easily removable stipe, fibrous,
very brittle in stipe.

<u>Basidiospores</u> brownish yellow-ochre in mass, 5.5-7(8.5) x
3.5-4.5 μm, elongate-elliptic in face-view, perhaps slightly
amygdaliform in side-view, smooth, non-amyloid, slightly
thick-walled, ochraceous in water; germ-pore absent. <u>Basidia</u>
4-spored, hyaline, clavate-pedicellate. <u>Hymenophoral trama</u>
subparallel. <u>Cheilocystidia</u> lecythiform, 16.5-21.5 x 6-7 μm,
capitulum 3.5-4 μm, some with brownish plasmatic pigment in
venter in KOH. <u>Pileipellis</u> a palisadoderm of pyriform to
ellipsoid-pedicellate cells some with slightly brownish
encrustations, 10-15 μm broad, some with plasmatic pigment,
hair-like cells absent; <u>pilocystidia</u> detached, either long
encrusted hyphae with clamp-connections (17.5-32 x (2.5-) 4.5-6 μm
or short chains arising from ± vesiculose cells c. 12 μm broad,
usually with swollen capitulum at apex and often containing
brown plasmatic pigment. <u>Pileus trama</u> of hyphae with yellow
contents in KOH. <u>Stipitipellis</u> of parallel to subparallel
cells producing lageniform to cylindric caulocystidia intermixed
with long flexuous hair-like cells <60 μm. <u>Clamp-connections</u>
present in pileus trama, pileipellis and stipitipellis.

<u>Material examined</u>: on soil and in litter, under tawa
(<u>Beilschmiedia tawa</u> Benth. & Hook. f.), <u>Cyathea</u>, <u>Hedycarya,</u>
<u>Schefflera</u> and <u>Alseuosmia</u>, Mt Kaitarakihi track, Coromandel,
8 v 1977, <u>Taylor</u> 1055.

The pilocystidia are distinctive and possibly represent
remnants of the original blematogen as described for <u>C. rickeniana</u>
(Singer) P D Orton and <u>C. mesospora</u> (Kühner ex) Kühner & Watling

(Watling, 1975). Similar structures have not been found previously
in C. piloselloides or C. pilosella (Pers.: Fr.) Kühner and may
indicate the New Zealand material represents an independent yet
closely related taxon. Strong developments of pilocystidia
are found in C. horakii Watling & Taylor (q.v.) and in C. dumetorum
(Vel.) Svreček.

The fungus is easily recognised by the small basidiospores
lacking a germ-pore, strongly developed often coloured pilocystidia
and lageniform to cylindric caulocystidia mixed with hair-like
cystidia. For these latter cells Watling (1964 et seq.) has
erroneously used the term pilocystidia, referring to the
dermatocystidia of the pileus as pileocystidia. Patrick & Barrows
(1979) are correct in pointing out that pilocystidia, a term
introduced by Buller is derived from Greek where pilo- refers to
head and not hair (Latin). This has led to some confusion; the
hair-like cells are so characteristic we believe they should be
recognised as a distinct entity.

7. C. pubescens (Gillet) Kühner, Le Genre Galera 85, 1935 (non
sensu Kühner, 1935). Fig. 16 E-I.
Basionym: Agaricus pubescens Gillet, Les Hymen. France 553,
1876.

No field data is available for the Waitakere Range collection
but the microscopic characters agree with the north temperate
agaric and are offered herein.

Basidiospores 15-16 x 8.5-9 µm, elliptic in face-view,
slightly flattened in side-view, thick-walled; germ-pore broad,
central. Basidia 24-26 x 11-13 µm. Cheilocystidia lecythiform,
16-20 x 8-10 µm with capitulum 3 µm broad. Pileipellis a
palisadoderm of spheropedunculate cells ⩾30 µm broad, intermixed
with hair-like cells ⩾75 µm long and ⩾2 µm broad, and tibiiform
cells 37 x 6 µm, some yellowish in alkaline solutions and with
capitulum >6 µm. Stipitipellis of cylindric, parallel hyphae
(7 µm broad) supporting mixture of hair-like cells with extremely
long neck and lecythiform cells 18-25 x 6-14 µm with head 3-4 µm
broad, but neck poorly differentiated. Clamp-connections
present.

Material examined: on horse dung, Matuku Reserve, Waitakere
Range, 29 x 1983, Taylor 1346.

One interesting feature of this collection is the blueing of
some of the hair-like cells on the pileus in alkali solutions.
Such a phenomenon has been found in neither European collections
of this taxon nor indeed, as yet, in other species of the genus
Conocybe. The significance of this unusual reaction must be
monitored. Bell (1983) records C. pubescens from dung in New
Zealand but there has apparently been some misinterpretation of
the data. The basidiospores of C. pubescens in Bell (as 'sensu
Richardson & Watling') is given as 8-9 x 5-6 µm when Richardson
& Watling (1968) give 'Basidiospores larger over 15 µm long x up
to 10 µm broad'; see Watling, 1982. With no further details
the species figured by Bell (1983) can neither be determined,
nor placed in either sect. Pilosellae or sect. Farinosae. We can
only think that the couplet in the key has been switched giving
C. pubescens the small spore-size, whilst C. rickenii was given
spores 17-20 x 10-13 µm.

8. C. rickenii (J. Schaeffer) Kühner, Le Genre Galera: 115,
1935.

Basionym: Galera rickenii J. Schaeffer in Zeits. f. Pilzk. 9:
171, 1930.

Recorded by Bell (1983) but, except for dimensions of the
basidiospores, without further data. The dimensions given viz.
17 - 20 x 10 - 13 µm are more in keeping with C. pubescens. The
spores should be (11-)12.5 - 16.5(- 17.5) x 7 - 10 µm for
C. rickenii (Watling, 1982). If we are correct in that a couplet
in the key has been switched the small spores (8- 9 x 5 - 6 µm)
still do not refer to C. rickenii. It is neither possible to
say to what the measurements refer nor whether the collection is
referable even to sect. Pilosellae. C. rickenii is characterised
by 2-spored basidia; a small spored taxon growing on dung in the
C. siliginea group is C. oculispora Locq.

9. C. rugosa (Peck) Watling in Beih. Nova Hedwigia 82: 133,
1981. Fig. 2 O-T; Fig. 14 J-N.

Basionym: _Pholiota rugosa_ Peck in Ann. Rep. New York State Mus. Nat.
Hist. 50: 102, 1897.

Description of New Zealand collection:

Pileus 27 mm, hemispheric, becoming convex to subcampanulate,
golden tawny brown, sienna or cinnamon fawn when fresh, strongly
hygrophanous, drying out whitish with ochraceous centre, dry,
smooth, membranaceous; margin striate. _Stipe_ 30-40 x 2.5 mm
(base 5 mm), cylindric, annulate, white then fawn, apex pruinose,
below ring longitudinally fibrillose or minutely squamulose, base
dark brown, dry, fragile, hollow; _ring_ mobile, whitish, persistent,
striate-grooved above, pale woolly below. _Gills_ adnexed to
adnate, ventricose, dense, concolorous with fresh pileus or
cinnamon to sienna, with whitish, minutely denticulate edge.
Flesh white. _Smell_ slight.

Basidiospores 8.5-9.5(-10.5) x 5-5.5 µm, elliptic in face-view
slightly flattened in side-view, thick-walled, non-amyloid;
germ-pore prominent, broad, central. _Basidia_ 4-spored, clavate
with short pedicel, 20-27 x 7 µm. _Cheilocystidia_ lageniform with
narrow tapered neck, 27-40 x 8-9 µm and obtuse apex 2-3 µm broad.
Caulocystidia similar to those on gill-edge 42-51 x 9-17 µm,
with apex 3-3.5 µm, intermixed with vesiculose cells, 23-32 x
9-13 µm. _Pileipellis_ a palisadoderm of spheropedunculate cells
27-36 x 15-17 µm.

Material examined: in potting soil, Landscape Road, Auckland,
2 iv 1980, _Segedin_ 1649; in soil under rotting log, rimu and
mixed broad-leaved trees, Old Coach Road, Waitakere Range,
20 iii 1983, _Taylor_ 1280.

This collection agrees in all ways with Peck's fungus as
interpreted by Watling (1982); see also van Waveren (1970). It
is widespread in North temperate areas often in sites associated
with human activity, e.g. in greenhouses and gardens. It is
probably an introduction to New Zealand.

10. _C._ aff. _vexans_ P D Orton in Trans. Brit. Mycol. Soc. 43:
197, 1960. fig. 14 A-I.

Description of New Zealand material:

Pileus 17-24 mm, hemispherical becoming convex to subcampanulate, brown, turning yellow-brown with age, strongly hygrophanous, smooth, membranaceous, dry; margin striate. Gills adnexed-adnate, ventricose, dense, pale (whitish), turning rust brown, with minutely dentate white edge. Stipe 32.5-37 x 1.5-2 mm (base < 3 mm), ±cylindric, annulate, apex pruinose, pale yellow to white, below ring longitudinally fibrillose or minutely squamulose, base dark brown, dry, hollow, fragile; ring mobile, whitish, striate-grooved, smooth beneath, persistent. Smell none.

Basidiospores 10.5-12.5 x 6-6.5 µm, elliptic in face-view, slightly flattened in side-view, thick-walled, ochraceous brown in water, slightly darker in aqueous alkali solutions, non-amyloid; germ-pore central, broad. Basidia 4-spored, clavate with short pedicel, hyaline in water and alkali solutions, 27.5-35 x 10-12.5 µm. Cheilocystidia ampulliform to cucurbitiform or lageniform 30-35 x 7.5-12 µm with distinct neck 7.5-15 x 2.5-3.5 µm, hyaline in water and alkali solutions; pleurocystidia absent. Pileipellis a strict palisadoderm of smooth, spheropedunculate cells 10-26 x 14-22 µm; pilocystidia absent. Stipitipellis of cylindric, hyaline, parallel hyphae, above the ring supporting groups of irregularly lageniform caulocystidia 35-42 x 8-12 µm with apex 3-4 µm, below ring with floccose hyphae. Clamp-connections present.

Material examined: on cow dung, near Buller bridge, west of Murchison, Nelson, 25 i 1969, Horak ZT 69/20.

This collection differs from C. vexans in its habitat preference; it is unknown on dung in Europe. Two coprophilous, annulate Conocybe spp. (sg. Pholiotina), occur in North America but they differ in their microscopic characters particularly the smaller basidiospores. C. vexans is the same as Kühner's four-spored form of C. blattaria; see Watling & Gregory (1981) and van Waveren (1970).

A second collection, Horak, in ZT 68/440 probably represents the same taxon although the pileus was wrinkled at the centre. We do not place much emphasis on this phenomenon which is not infrequent in many members of the Bolbitiaceae; compare C. rugosa and C. filaris in Watling (1982). This collection, from under Leptospermum ericoides A. Rich., Wharariki, Cape Farewell, Nelson, 13 v 1968, has both slightly narrower, and

therefore slightly fusoid spores and narrower cheilocystidia;
the following description is offered:

Pileus 14-17 mm, umbonate-convex to campanulate, dark ochre-
brown (wet), ochraceous (dry), hygrophanous, micaceous-shiny,
at least centre wrinkled-venose, dry, striate; veil remnants
absent. Gills adnate to adnexed, ventricose, pale ochre (young),
chocolate brown (with rust brown tinge) when old, crowded, with
white, fimbriate edge. Stipe 25-34 x 1-1.5 mm (base 2 mm)
cylindrical or subclavate, annulate, concolorous with pileus,
subpruinose at apex, fibrillose towards base, hollow, dry;
ring persistent, striate-grooved, immobile. Smell and taste
not distinctive.

Basidiospores 10-12.5 x 5.5-6.5 µm, ellipsoid in face-view,
slightly flattened in side-view, thick-walled, ochraceous brown
in water, slightly darker in aqueous alkali solutions, non-
amyloid; germ-pore central, broad. Basidia 4-spored, 24-25 x
5-9 µm. Cheilocystidia ampulliform to cucurbitiform with
distinct elongate neck, 30-37 x 8-11 µm with apex 2 µm broad.
Pileipellis a palisadoderm of pedicellate cells, 22-40 x 12-22 µm.

Conocybe spp.

At least four other members of the genus are known from New
Zealand but are inadequately documented as only single basidiomata were apparently collected. For completeness these
collections are added.

11. Conocybe sp. 1 Fig. 16 A, R & S

Pileus 14 mm, warm fawn with mahogany disc, dull, non-striate;
margin very scalloped. Stipe 54 x 25 mm, faintly tapering
towards base, whitish at base, paler than pileus, darkening
downwards with faint silky sheen, solid. Gills adnexed,
mustard-yellow near stipe, to mahogany near margin of pileus.
Flesh yellowish in pileus, thin especially at margin, yellowish
in stipe.

Basidiospores brownish yellow in mass, 15.5-17.5 x 8.5-9.5 µm,
medium ochraceous in water, deep brown in aqueous alkali
solutions; germ-pore broad, prominent. Basidia 2-spored, 19-28.5

x 9.5-10.5 µm becoming brownish yellow with age. <u>Cheilocystidia</u>
lecythiform, 15.5-24 x 7-9.5 µm, capitulum 3-4 µm. <u>Caulocystidia</u>
lecythiform 15.5-18 x 8-9.5 µm, capitulum 2.5-4 µm, with a few
hair-like cells, intermixed at stipe-apex with 2-spored basidia.
<u>Pileipellis</u> a palisadoderm of spheropedunculate cells, 25.5-38 µm
broad, intermixed with a few lecythiform dermatocystidia.

<u>Material examined</u>: In pasture at edge of forest, hill above
pylon, Wiltons Bush, Wellington, 3 iv 1961, legit L.R. Taylor,
<u>Taylor</u> 64. This collection should be compared with <u>C. rubiginosa</u>
Watling with similar microscopic details but <u>Taylor</u> 64 was of a
rather desiccated specimen and conspecificity could not be
confirmed. Unfortunately as shown by van Waveren (1970) and
Watling (1971) the pileus once dried never returns to its original
colour although it might darken.

12. <u>Conocybe</u> sp. 2

The mixed collection of Colenso's <u>b</u> 113 contains a single
basidioma agreeing very closely with the fungus described above
as <u>Conocybe</u> sp. 1.

<u>Basidiospores</u> 15-16.5 x 8-9.5 µm, ellipsoid, red-brown in
water, smooth, thick-walled; germ-pore prominent, broad. <u>Basidia</u>
2-spored. <u>Cheilocystidia</u> collapsed, capitulum 4-4.5 µm broad.
<u>Caulocystidia</u> mixture of lecythiform and hair-like cells.

Material examined: as <u>Agaricus pediades</u>, Colenso <u>b</u> 113 (K).

<u>Galera tenera</u> Schaeffer ex Fries (= <u>Conocybe tenera</u> (Schaeffer
ex Fries) Fayod) has been recorded by Colenso (1890) and by
Massee (1898). Unfortunately it is not possible to confirm early
records as most small brown '<u>Galera</u>'-like agarics resembling
<u>Agaricus tener</u> were lumped under that epithet; see <u>Psilocybe</u>
pg. 54.

13. <u>Conocybe</u> sp. 3 Fig. 16 J-M, P & Q.

<u>Pileus</u> 'small', brown when moist, hygrophanous, drying
ochraceous tan with dark band around margin. <u>Stipe</u> hoary, with
longitudinal brown lines. <u>Gills</u> almost free, clay brown.

<u>Basidiospores</u> lenticular, hexagonal to angled, broadly
elliptic in face-view, flattened in side-view, 8-9(-10) x (5.5)

- 51 -

6-6.5 µm, x 5.5-6 µm, yellow brown in water and alkali solutions,
thick-walled; germ-pore broad, central. <u>Basidia</u> cylindrico-
clavate, 29-34 x 7-9 µm with foot 2.5-3 µm broad and basal clamp-
connections. <u>Cheilocystidia</u> lecythiform, 13-17 x 5-7 µm,
capitulum 3-5 µm broad. <u>Pileipellis</u> a palisadoderm of
spheropedunculate cells 42 µm broad, some with encrusted
pedicel, others with yellow vacuolar pigmentation, and intermixed
with capitate cells, narrower than cheilocystidia, 19-33 x 3-6 µm
with head ≤4 µm, many filled with persistent yellow plasmatic
pigment. <u>Stipitipellis</u> of clamp-connected, parallel ± encrusted
hyphae, with ± yellow plasmatic pigment dissolving in alkali
solutions; <u>caulocystidia</u> similar in shape, or more variable
than pilocystidia. <u>Clamp-connections</u> present.

   <u>Material examined</u>: on dung (cow)?, Pukekawa?, 19 v 1963,
legit R.S. Lediard, <u>Segedin</u> 219.

   This is obviously a distinct taxon but further collections
are required to expand the field data. This agaric on spore-
shape alone is close to the European <u>C. antipus</u> (Lasch) Fayod
but lacks the elongation of the stipe into a long pseudorhiza.
<u>C. lenticulospora</u> Watling, which also grows on dung and has
lenticular basidiospores, is a member of sect. <u>Pilosellae</u> whereas
<u>Segedin</u> 219 is assignable to sect. <u>Conocybe</u>. <u>C. hexagonospora</u>
Métrod nomen nudum has similar sized spores but no type material can
be traced (Heim, in litt.); it is reported as having some grey
tints, and grew on burnt ground: see Bell's <u>C. pubescens</u> (1983)
discussed on pg. 47.

14. <u>Conocybe</u> sp. 4 Fig. 16 B-D.
   <u>Pileus</u> 24 mm, fawn brown, not shiny, convex-campanulate.
<u>Stipe</u> 62 x 2 mm cartilaginous, twisted, equal except for slightly
bulbous base. <u>Gills</u> free-adnexed, cinnamon. <u>Smell</u> strongly
nitrous.

   <u>Basidiospores</u> 13.5-16 x 9.5-12 x 8.5-9 µm, broadly elliptic
to ovate in face-view, distinctly flattened in side-view, thick-
walled; germ-pore broad, central. <u>Basidia</u> 4-spored.
<u>Cheilocystidia</u> lecythiform, 16-19 x 9-10.5 µm, capitulum 3-4 µm
broad. <u>Pileipellis</u> a palisadoderm with spheropedunculate cells

42-44 μm broad, infrequently intermixed with lecythiform cells.
Clamp-connections present; oleiferous hyphae present in pileus.

Material examined: incubated in petri-dish; on dung,
Titirangi, 3 i 1963, legit R.S. Lediard, Segedin 165.

Unfortunately no caulocystidia were located; it was therefore
impossible to assign the collection to a particular section,
although it undoubtedly belongs to sg. Conocybe. The large size
of the basidiospores and the poor differentiation of the neck of
the cheilocystidia suggest a member of the C. pubescens (Gillet)
Kühner group; the nitrous smell is unusual.

Rejected and Misidentified collections

Cortinariaceae

Descolea

Basidiomata small to medium, pholiotoid, never 'deliquescent',
uniformly rust, tawny or of similar bright shades of brown.
Pileus campanulate to convex, unexpanding or becoming plano-convex,
hygrophanous or expallent.

Three species of this genus are known from New Zealand:-
D. majestatica Horak, found in South Island with Nothofagus spp.,
D. phlebophora Horak, also found with Nothofagus but in
addition with Leptospermum, and D. gunnii (Berk.) Horak based on
Secotium gunnii (Gunn 257 in K) noted by Berkeley (in Massee, 1891)
and originally collected at Rotorua. The last species grows with
both Leptospermum spp. and Nothofagus spp. and is fairly common
in coastal and submontane forests; it is also known from New
Guinea (Horak, 1980d). D. gunnii is closely related to the
Australian D. recedens (Cooke & Massee) Singer now known to be
widespread in Eastern Australia (Horak, 1980d; Watling unpubl.
data). Descolea has been recently placed in the Bolbitiaceae by
Singer (1972) and Horak (1979b).

Galerina sp. 1

as Agaricus vervacti xii 1885 Colenso b 71.

We agree with Horak (1971b) in that this collection of
A. vervacti belongs to Galerina; see Agrocybe vervacti, pg. 26.

<u>Galerina</u> sp. 2

as <u>Agaricus</u> (Naucoria) <u>pediades</u> in field, Colenso 269.

This collection belongs in the genus <u>Galerina</u> and agrees with the illustration for Colenso 1039 (<u>Naucoria nasuta</u>) which appears in Horak (1971b); see below.

<u>Galerina</u> sp. 3

as <u>Agaricus (Naucoria) pediades</u>, under Kaiwarawarra, T. Kirk (sheet H1257/80 in K). Resembling in all ways the <u>Galerina</u> figured under <u>N. nasuta</u> Kalchbrenner (Horak, 1971b). The basidiospores are very distinctive in that they possess a roughened calyptrospore; the pleurocystidia usually possess two lobes at the apex.

Lepiotaceae

<u>Leucoagaricus holosericeus</u> (Fries) Moser apud Gams

as <u>Agaricus praecox</u>, in field, Napier, Colenso, <u>b</u> 283 and retained there by Horak (1971b) but we differ in our interpretation of the material. The basidiospores are metachromatic, dextrinoid (9.5–10.5 x 6 µm) and the cheilocystidia characterise a member of the <u>L. naucina</u> group. Parallel species are known from New Zealand and they may well have been introduced; they are generally more frequent in countries with warmer climates. A collection has been located in <u>PDD</u> (40317) filed under <u>Armillaria</u>.

Strophariaceae

<u>Psilocybe</u>

<u>Psilocybe</u> sp. 1.

as <u>Galera tenera</u>, Waitaki, <u>Berggren</u> 60 (K); Berggren material is the subject of a paper by Cooke in Grevillea (1890).

Only the following microscopic data was obtainable:-
<u>Chrysocystidia</u> absent. <u>Basidiospores</u> 16.5–18 x 9.5–10.5 µm, mid-brown ochraceous with olivaceous tinge, smooth, thick-walled, sometimes with dark inclusions; germ-pore prominent, <u>Pileipellis</u> filamentous. Neither cheilo- nor pleurocystidia could be located.

We agree with Horak (1971b) in placing this collection in <u>Psilocybe</u>. It is impossible to be any more precise even using

Guzmán's monograph (1983).

Psilocybe sp. 2

as _Agaricus semiorbicularis_, Waitaki, _Berggren_ 61 (K).
Badly consumed by insects.

Only the following microscopic data was obtainable:
Cheilocystidia absent. _Basidiospores_ 11-12 x 6.5-7(-7.5) µm,
smooth, thick-walled with truncate germ-pore, sometimes with
dark inclusions. _Pileipellis_ filamentous. Neither cheilocystidia
nor pleurocystidia could be found.

This belongs to _Psilocybe_; see comments above. We do not
agree with the placement by Horak (1971b).

An illustration in E apparently is a copy by M.C. Cooke of a
Berggren collection labelled 'AGARICUS (GALERA), on the ground,
New Zealand'. It is impossible to go any further but it is not
unlikely that this figure depicts one of the above collections
of _Psilocybe_.

Stropharia

Stropharia coronilla (Bull.: Fr.) Quel.

This species can easily be confused with members of the genus
Agrocybe and the _Agaricus semotus_ group. It is characterised by
purple-brown basidiospores with germ-pore, and distinct
chrysocystidia. Probably introduced into New Zealand (in grass
of orchard, Oamaru, 22 i 1966, legit L.R. Taylor, _Taylor_ 248).

Tricholomataceae

as _Agaricus temulentus_, Colenso _b_ 751

Indicated by Horak (1971b) to lack all the essential features
of the Bolbitiaceae. The basidiospores are 8-10 x 5-6 µm,
kidney-shaped, smooth and honey-coloured in ammoniacal solutions.
The cheilocystidia are diverticulate and pleurocystidia absent.
There is controversy concerning the true identity of _Ag._
temulentus. Thus material in Kew collected at Glamis, Scotland
by the Rev. J. Stevenson is _Agrocybe arvalis_ (Fr.) Singer, yet
material identified by M.C. Cooke and cited by Orton (in Dennis,
Orton & Hora, 1960) is in fact a member of the Tricholomataceae
(see also Reid & Austwick, 1963).

ACKNOWLEDGEMENTS

The authors are indebted to Dr Egon Horak, Zürich and Professor E J H Corner, Cambridge for loan of material and to the former for making available his field and microscopic observations. Collections and notes were also lent by Dr B Segedin, Auckland and Dr P Austwick, London. Help has been given by Mrs Dorothy Brunton in the preparation of the plates based on several sources of observation some our own and others of Egon Horak. Mrs N Gregory and Miss H Kronast have been of great assistance in the final preparation of the manuscript.

One of us (MT) wishes to thank the Regius Keeper, Royal Botanic Garden, Edinburgh for allowing her to spend a 5 month study period in the Garden laboratory. She also thanks the University of Auckland for granting her leave-of-absence over the same period.

REFERENCES

BELL, A. (1983). _Dung Fungi: an illustrated guide to coprophilous fungi in New Zealand_, Wellington.

BERKELEY, M.J. (1855). In Hooker: _Flora Novae-Zelandiae_ Vol. 2, _Flowerless Plants_, London

_____ (1891). In Massee, _Grevillea_ 19: 94-98.

COLENSO, W. (1886). An enumeration of fungi recently discovered in New Zealand with brief notes on the species novae. _Trans. Proc. N.Z. Inst._ 19: 301-313.

_____ (1890). ditto. _Trans. Proc. N.Z. Inst._ 23: 391-398.

COOKE, M C. (1890). New Zealand fungi. _Grevillea_ 19: 1-4.

_____ (1892). _Handbook of Australian Fungi_, London.

CUNNINGHAM, G H. (1942). _The Gasteromycetes of Australia and New Zealand_, Dunedin.

_____ (1963). The Thelephoraceae of Australia and New Zealand, _N.Z. Dept. Sci. Industr. Res. Bull._ 145, Wellington.

_____ (1965). Polyporaceae of New Zealand _N.Z. Dept. Sci. Industr. Res. Bull._ 164, Wellington.

DENNIS, R W G. (1953). Les Agaricales de l'Ile de la Trinité: Rhodosporae-Ochrosporae. _Bull. Soc. Mycol. Fr._ 69: 145-198.

_____, ORTON, P D & HORA, F B (1960). New checklist of British agarics and boleti. _Trans. Brit. Mycol. Soc._ 43. suppl.

ESSER, K, SEMERDZIEVA, M & STAHL, U (1974). Genetische Untersuchungen an dem Basidiomyceten _Agrocybe aegerita._ Theor. & Appl. Genetics 45: 77-85.

ESSER, K & MEINHARDT, F. (1977). A common genetic control of dikaryotic and monokaryotic fruiting in the Basidiomycete _Agrocybe aegerita._ _Molec. gen. Genet._ 115: 113-115.

FERRI, F. (1973). La coltivazione del pioppino (_Agrocybe aegerita_ (Brig.) Singer). _Micol. Ital._ 2: 29-33.

FUHRER, B. (1985) A Field Companion to Australian Fungi. Melbourne, Victoria

GUZMÁN, G. (1983). _The genus Psilocybe._ _Beih. Nova Hedw._ 74: 1-439.

HEIM, R. (1968). Deuxième mémoire sur les Cyttarophyllés. _Bull. Soc. Mycol. Fr._ 84: 103-116.

HEINEMANN, P. (1974). Quelques _Agaricus_ de Nouvelle-Zélande.
Bull. Jard. Bot. Nat. Belg. 44: 355-366.

HENDERSON, D M, ORTON, P D & WATLING, R. (1969). _The British_
Fungus Flora: Agarics & Boleti. Introduction, (_An_
Identification Colour Chart: Flora of British Fungi),
Edinburgh.

HOOKER, J D. (1855). _The Botany of the Antarctic Voyage:_
II. _Flora Novae-Zelandiae_, London.

HORAK, E. (1971a). Studies in the genus _Descolea_ Singer.
Persoonia 6: 231-248.

_______ (1971b). A contribution towards the revision of the
Agaricales (Fungi) from New Zealand. _N.Z. J. Bot._ 9: 403-462.

_______ (1971c). Contribution to the knowledge of the Agaricales s.l.
(Fungi) of New Zealand. _N.Z. J. Bot._ 9: 463-493.

_______ 1973a). Fungi Agaricum Novazelandiae I. _Entoloma_ (Fr.)
and related genera. _Beih. Nova. Hedw._ 43: 1-86.

_______ (1973b). ibid II. _Thaxterogaster_ Singer (1951). _Beih._
Nova. Hedw. 43: 87-114.

_______ (1973c). ibid III. _Hygrophorus_ Fr. and related genera.
Beih. Nova. Hedw. 43: 115-182.

_______ 91973d). ibid IV. _Phaeocollybia_ Heim (1931). _Beih._
Nova Hedw. 43: 183-192.

_______ (1973e). ibid V. _Cuphocybe_ Heim (1951). _Beih. Nova._
Hedw. 43: 193-200.

_______ (1977). ibid _VI. Inocybe_ (Fr.) Fr. and _Astrosporina_
Schroeter. _N.Z. J. Bot._ 15: 713-747.

_______ (1979a). ibid VII. _Rhodocybe_ Maire. _N.Z. J. Bot._ 17: 251-281

_______ (1979b). Agaricales y Gasteromycetes secotioides. _Fl._
Cript. Tierra del Fuego 11: 1-525.

_______ (1980a). ibid VIII. _Phaeomarasmius_ Scherrfel and
Flammulaster Earle. _N.Z. J. Bot._ 18: 173-182.

_______ (1980b). ibid IX. _Lepiotula_ (Maire) Locquin ex Horak.
N.Z. J. Bot. 18: 183-188.

_______ (1980c). ibid X. _Simocybe_ Karsten. _N.Z. J. Bot._ 18: 189
-196.

_______ (1980d). New and Remarkable Hymenomycetes from Tropical
Forests in Indonesia (Java) and Australasia. _Sydowia_ 33:
39-63.

KÜHNER, R. (1935). <u>Le Genre Galera</u>, Paris.

MACDONALD, R. & WESTERMAN, J. (1979). Fungi of South-eastern Australia. Melbourne.

MACNABB, R.F.R. (1967). Strobilomycetaceae of New Zealand. <u>N.Z. J. Bot</u>. 5: 532-547.

______ (1968). Boletaceae of New Zealand. <u>N.Z. J. Bot</u>. 6: 137-176.

______ (1971). The Russulaceae of New Zealand, 1. <u>Lactarius</u> DC. ex S F Gray. <u>N.Z. J. Bot</u>. 36: 41-59.

______ (1972). Tricholomataceae of New Zealand 1. ·<u>Laccaria</u>. <u>N.Z. J. Bot</u>. 10: 461-484.

______ (1973). The Russulaceae of New Zealand, 2. <u>Russula</u> Pers. ex. S F Gray. <u>N.Z. J. Bot</u>. 11: 673-730.

MASSEE, G. (1891). New or imperfectly known Gastromycetes. <u>Grevillea</u> 19: 94-97.

______ (1898). The Fungus Flora of New Zealand. <u>Proc. N.Z. Inst</u>. 31: 282-349.

MEINHARDT, F. & ESSER, K. (1981). Genetic Studies of the Basidiomycete <u>Agrocybe aegerita</u>. <u>Theor. Appl. Genet</u>. 60: 265-268.

OVERHOLTS (1927). A monograph of the genus <u>Pholiota</u>. <u>Ann. Miss</u>. <u>Bot. Gdn</u>. 14: 87-210.

REID, D.A. (1986) New or interesting records of Australasian basidiomycetes VI. <u>Trans. Brit. Mycol. Soc</u>. 86: 429-440.

______ & AUSTWICK, P. (1963). An annotated list of the less common Scottish basidiomycetes (exclusive of rusts and smuts). <u>Glasgow Nat</u>. 18: 255-336.

REIJNDERS, A.F.M. (1971). The veil of <u>Agrocybe aegerita</u>. <u>Acta Bot. Neerl</u>. 20: 299-304.

RICHARDSON, M.J. & WATLING, R. (1968). <u>Keys to Fungi on Dung</u>, Cambridge.

PATRICK, W.W. & BARROWS, C. (1979). Western fungi: a New Mexico <u>Psathyrella</u> in the <u>Cystidiosae</u> subgenus <u>Homophron</u>. <u>Mycoxtaxon</u> 9: 493-500.

SINGER, R. (1950). Type studies on Basidiomycetes. <u>Lilloa</u> 23: 145-246.

______ (1951). <u>The Agaricales in Modern Taxonomy, Lilloa</u> 22: 1-832.

______ (1962) <u>The Agaricales in Modern Taxonomy</u>, 2nd edition, Weinheim.

SINGER, R. (1969). Mycoflora Australis, <u>Beih.Nova Hedw.</u>
28: 7-405.

_____ (1972). <u>The Agaricales in Modern Taxonomy</u>, 3rd.
edition, Vaduz.

_____ (1978). Keys for the Identification of the species of
Agaricales. <u>Sydowia</u> 30: 192-279.

STEVENSON, G. (1962a). The Agaricales of New Zealand II. <u>Kew
Bull.</u> 16: 65-74.

_____ (1962b). ibid III. <u>Kew Bull.</u> 16: 227-237.

_____ (1962c). ibid IV. <u>Kew Bull.</u> 16: 373-384.

_____ (1964). ibid V. <u>Kew Bull.</u> 19: 1-59.

_____ (1982a). <u>Field Guide to Fungi</u>. Christchurch.

_____ (1982b). A Parasitic member of the Bolbitiaceae. <u>New Zealand
J. For.</u> 27: 130-133.

TAYLOR, G M. (1968). Some New Zealand Mushrooms, <u>Tuatara</u> 16:
123-126.

_____ (1970). <u>Mushrooms and Toadstools in New Zealand</u>, Wellington.

_____ (1981). <u>Mushrooms and Toadstools</u>. Mobil New Zealand Nature
Series, Wellington.

_____ (1983). Some Common Fungi of Auckland City. <u>Tane</u> 29:
133-141.

WATLING, R. (1964). The taxonomic characters of the Bolbitiaceae
with particular references to the genus <u>Conocybe</u>. Ph.D. thesis,
Edinburgh.

_____ (1965). Observations on the Bolbitiaceae II. Conspectus
of the family. <u>Notes Roy. Bot. Gdn. Edinb.</u> 26: 289-323.

_____ (1971). Observations V. Developmental studies on <u>Conocybe</u>
with particular reference to the annulate species. <u>Persoonia</u>
6: 281-289.

_____ (1975). Studies on fruit-body development in the Bolbitiaceae
and the implications of such work. <u>Beih. Nova Hedw.</u> 51:
319-346.

WATLING, R. (1976). Observations XV. The taxonomic
position of those species of Conocybe with ornamented
basidiospores. Rev. Mycol. Paris, NS 40: 31-37.

_____ (1982). British Fungus Flora: Agarics and Boleti 3.
Bolbitiaceae, Edinburgh.

_____ (1983). Observations XXII. Further validations.
Notes Roy. Bot. Gdn. Edinb. 40: 537-558.

_____ (1985). Observations XXV. Icelandic species of
Bolbitiaceae. Acta Bot.Isl. 8: 3-19.

_____ & ABRAHAM, S.P. (1986). Observations 26.
Bolbitiaceae of Kashmir with particular reference to
the genus Agrocybe. Nov. Hedw. 42: 387-415.

_____ & GREGORY, N. M. (1980). Larger Fungi from Kashmir.
Nova. Hedw. 32: 493-564.

_____ & GREGORY, N.M. (1981). Census catalogue of world
members of the Bolbitiaceae. Biblio. Mycol. 82:1-224.

_____ QUADRACCIA, L., TABARÉS, M. & ROCABRUCA, A. (1986).
Gastrocybe in Europe. Notes Roy. Bot. Gdn.,Edinb.
43: 307-311.

WAVEREN, Kits E van. (1970). The genus Conocybe subg.
Pholiotina 1. European annulate species.
Persoonia 6:119-165.

ABSTRACT

A summary of all species attributable to the three
genera Agrocybe, Bolbitius and Conocybe and recorded
for New Zealand is given.   Thirteen species of Agrocybe,
three of Bolbitius, and nine species of Conocybe
are considered, three of which are new to science
viz. C. gracilenta, C. horakii and C. novaezelandiae.
A further species Agrocybe olivacea is described as new
from New Zealand.   The variation in members of A. cylindrica
complex is discussed.   Wherever possible the results from the
examination of European material has been called upon to offer
opinions as to whether the taxa are introduced or endemic
to New Zealand.

Fig. 1. AGROCYBE

A. puiggarii: A-F <u>Segedin</u> 760, A. Habit sketch;
B. Pilocystidia; C. Caulocystidia; D. Pleurocystidia;
E. Basidiospores; F. Cheilocystidia. G-J <u>Segedin</u> 1723,
G. Basidiospores; H. Basidia; I. Pleurocystidia;
J. Cheilocystidia. O-S <u>Segedin</u> 1334, O. Caulocystidia;
P. Pleurocystidia; Q. Cheilocystidia; R. Basidiospores;
S. Pilocystidia.
A. cf. acericola: K-N <u>Segedin</u> 1631, K. Pleurocystidia;
L. Cheilocystidia; M. Basidiospores; N. Caulocystidia.

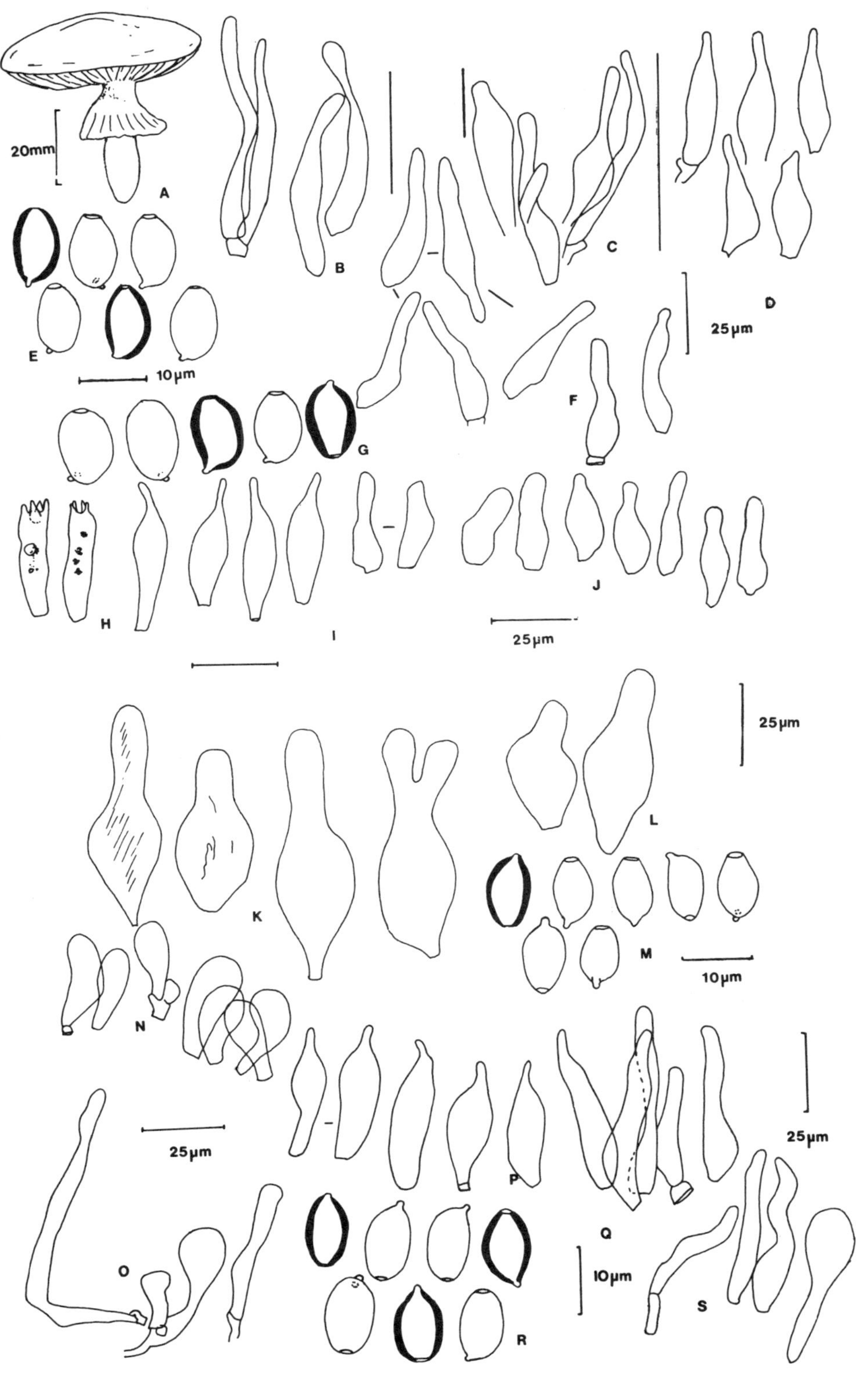

20mm
A
E
10µm
G
F
25µm
D
H
I
25µm
J
25µm
K
L
25µm
M
10µm
N
O
25µm
P
Q
R
10µm
S
25µm

Fig. 2. AGROCYBE AND CONOCYBE

A. howeana: A. Pleurocystidia PDD 29104;
B. Cheilocystidia PDD 29104;   D. Cheilocystidia
PDD 680;   E. Pleurocystidia PDD 680.
A. praecox agg: C. Cheilocystidia PDD 29208.
C. horakii: F-K Horak 68/66 holotype,
F. Pileipellis; G. Caulocystidia; H. Cheilocystidia;
I. Basidia; J. Basidiospores; K. Habit sketch and
section.   L-N Horak 69/24, L. Caulocystidia;
M. Cheilocystidia; N. Basidiospores.
C. rugosa: O-T Segedin 1649, O. Habit sketch;
P. Cheilocystidia, Q. Pileipellis units; R. Caulocystidia;
S. Basidiospores; T. LS. Section through hymenophoral
trama.

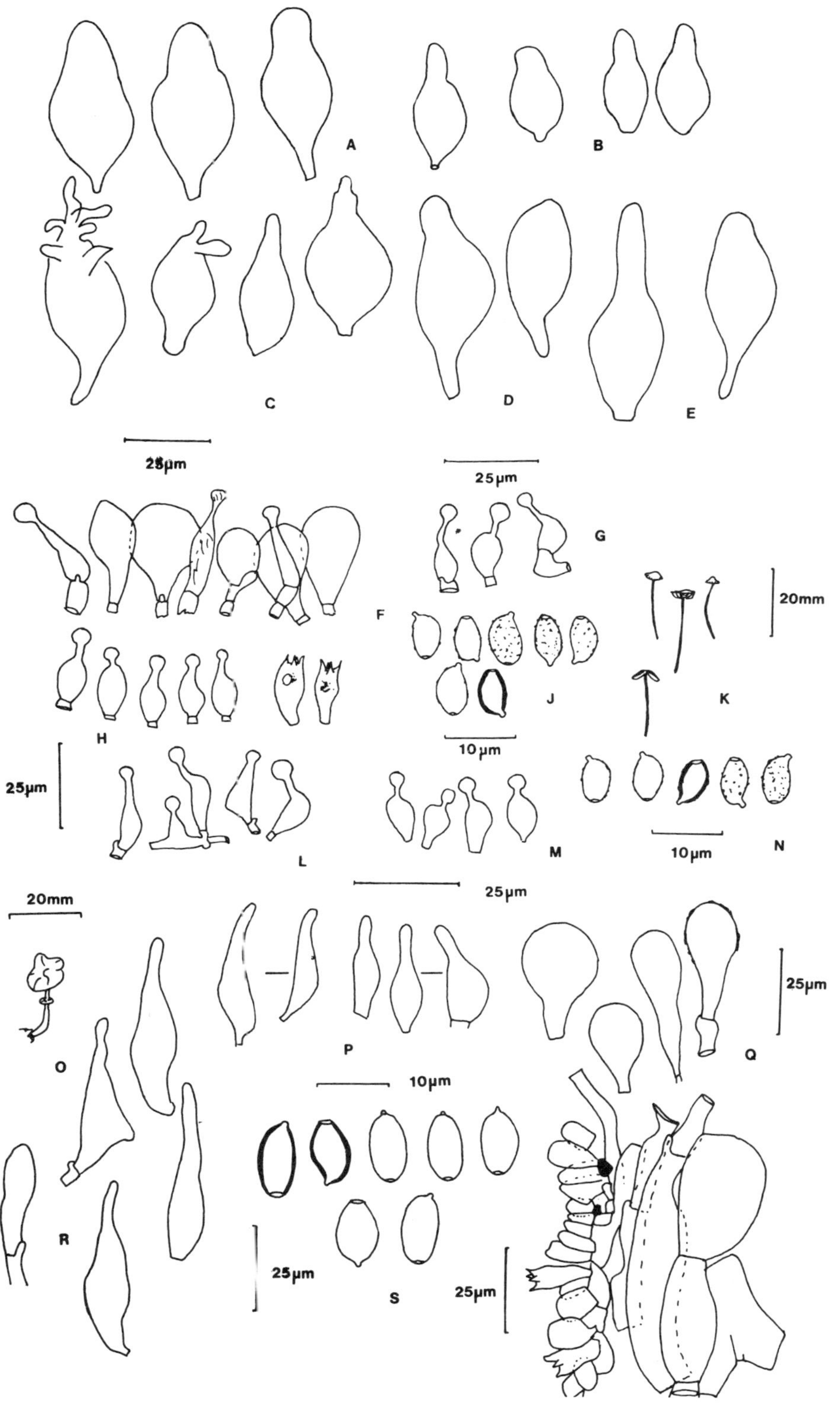

Fig. 3. AGROCYBE AND CONOCYBE

A. olivacea holotype: A. Caulocystidia;

B. Pleurocystidia; C. Cheilocystidia;

D. Basidiospores; E. Basidia; N. Habit sketch.

A. semiorbicularis (as pediades): F-I

Colenso ex K, F. Basidiospores; G. Basidia;

H. Caulocystidia; I. Cheilocystidia.

C. huijsmanii: J-P Taylor 1151,

J. Basidiospores; K. Basidium; L. Cheilocystidia;

M. Pileipellis; O. Habit sketch Taylor 1151 (AK1)

P. Habit sketch Taylor 1151 (AK2).

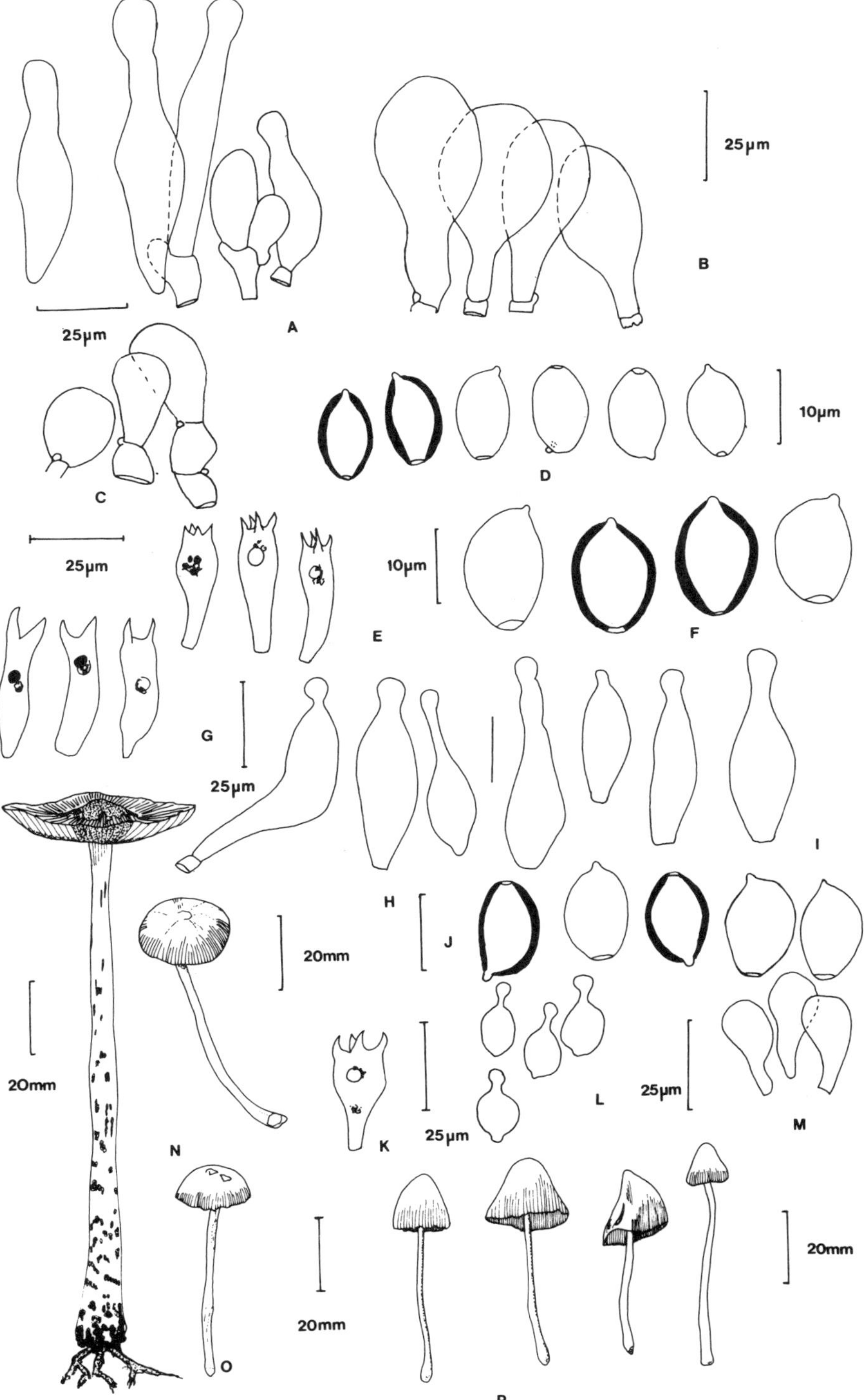

Fig. 4. AGROCYBE

<u>A. parasitica</u>: A. Cluster of 3 basidiomata and section (B), <u>Taylor</u> 702; C. Basidioma with primordium and section (D), <u>Taylor</u> 50; E. Section and basidioma, <u>Taylor</u> 52; F. Cluster of 2 basidiomata and section (G), <u>Taylor</u> 867.

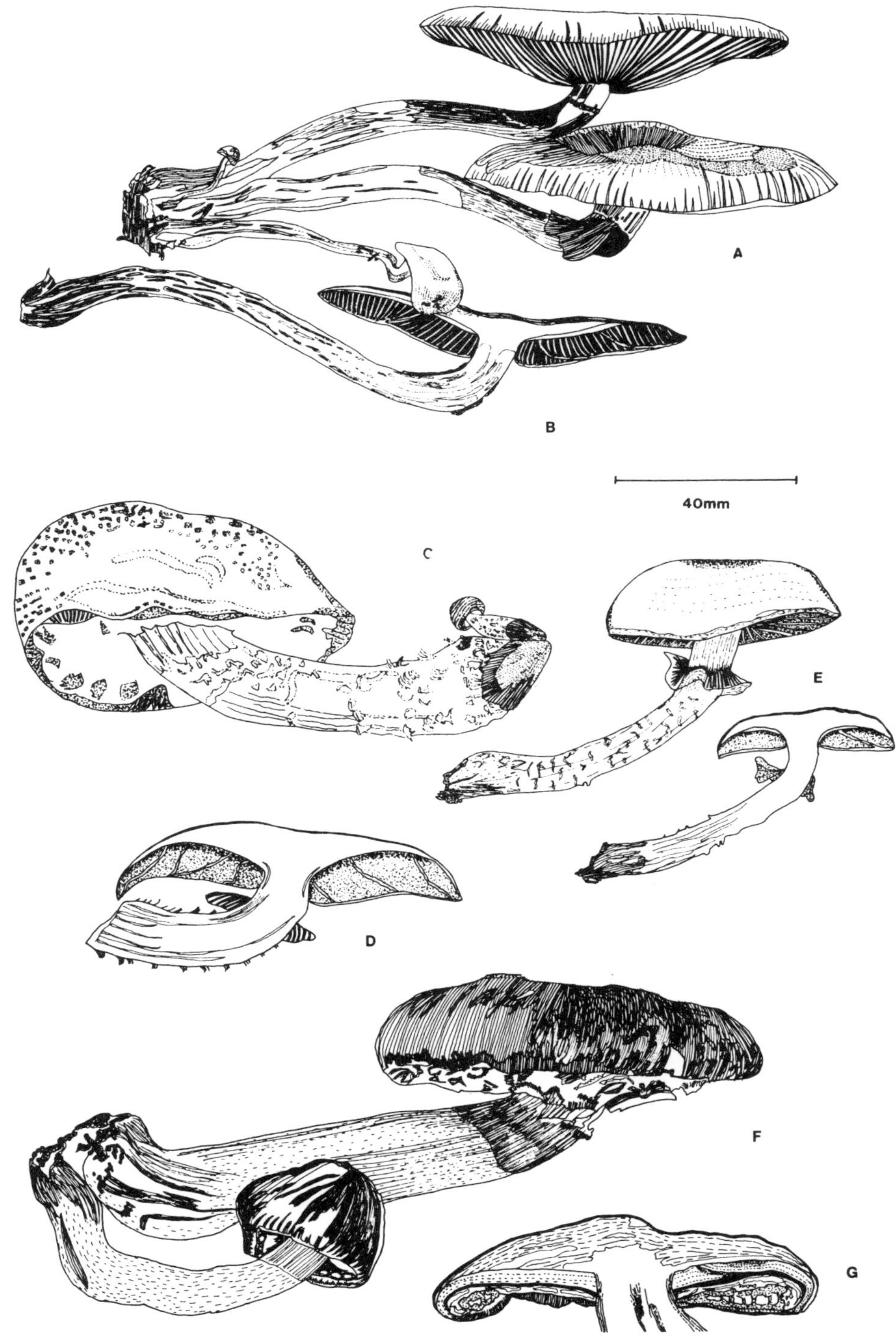

A

B

C

D

E

F

G

Fig. 5. AGROCYBE

<u>Agrocybe</u> <u>parasitica:</u> A-C <u>Taylor</u> 1365, A. Cheilocystidia;
B. Pleurocystidia; C. Basidiospores; D-I <u>Segedin</u> 1732,
D. Pileipellis;   E. Pilocystidium;   F. Basidiospores;
G. Cheilocystidia; H. Pleurocystidia; I. Basidia;
J-N <u>Segedin</u> 1564, J. Basidiospores; K & N Caulocystidia;
L. Pleurocystidia; M. Cheilocystidia; O-S <u>Segedin</u> 1734,
O. Pileipellis; P. Basidiospores; Q. Hyphae from below
ring; R. Hyphae from above ring; S. Hyphae of ring.

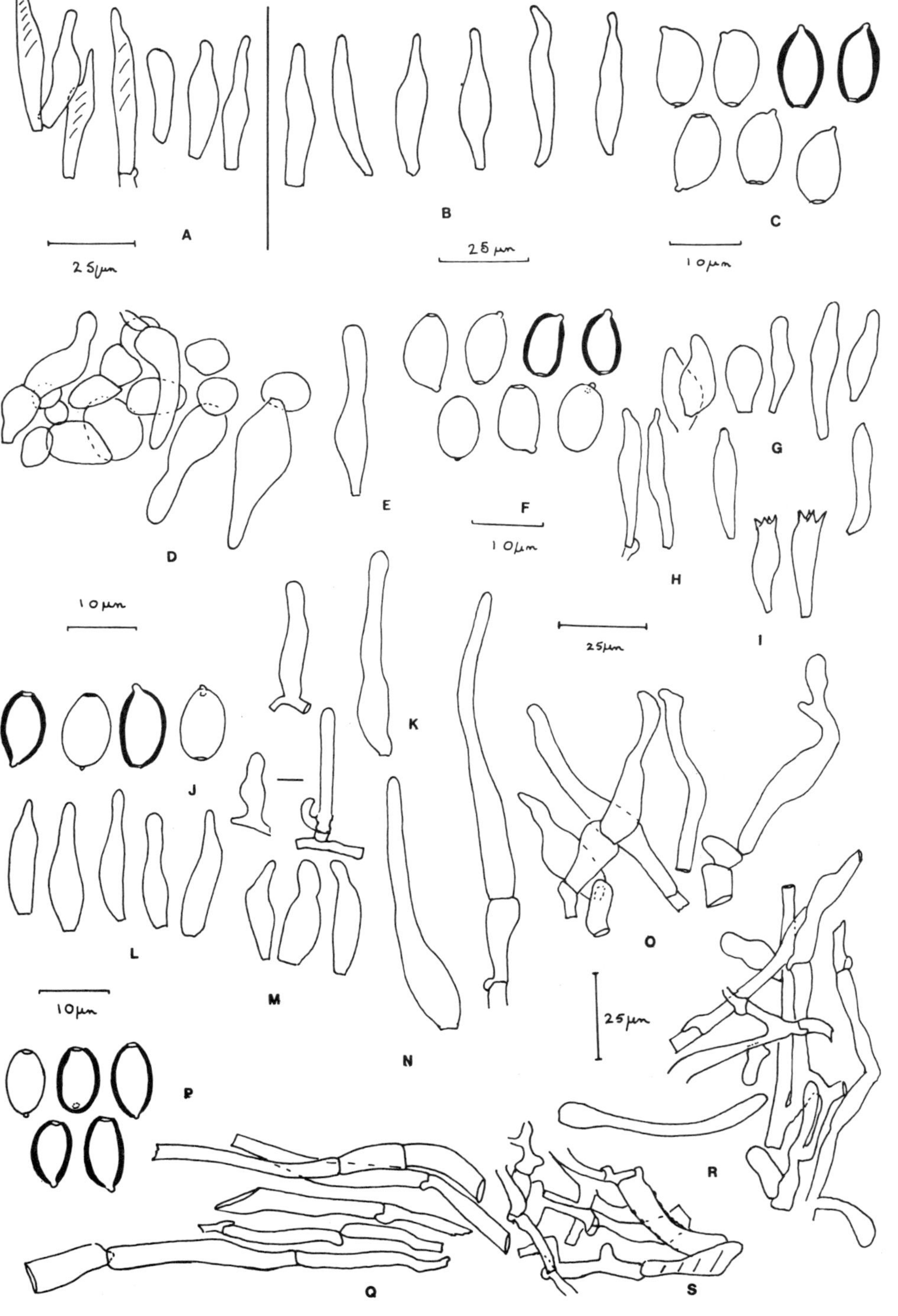

Fig. 6. AGROCYBE

<u>Agrocybe</u> <u>parasitica</u>: A-D <u>Taylor</u> 1370, A. Cheilocystidia;
B. Basidiospores; C. Pleurocystidia: D. Basidium;
E-F <u>Taylor</u> 1374, E. Basidiospores; F. Cheilocystidia;
G-H <u>Taylor</u> 702, G. Cheilocystidia; H. Pleurocystidia.
I. Pleurocystidia <u>Taylor</u> 52. J & N <u>Taylor</u> 50
J. Cheilocystidia; N. Pleurocystidia. K-M <u>Segedin</u> 894,
K. Pileipellis; L. Velar remnants on gill edge;
M. Pleurocystidia. O. Pleurocystidia <u>Taylor</u> 867.
P-Q <u>Segedin</u> 894, P. Pleurocystidia; Q. Cheilocystidia.

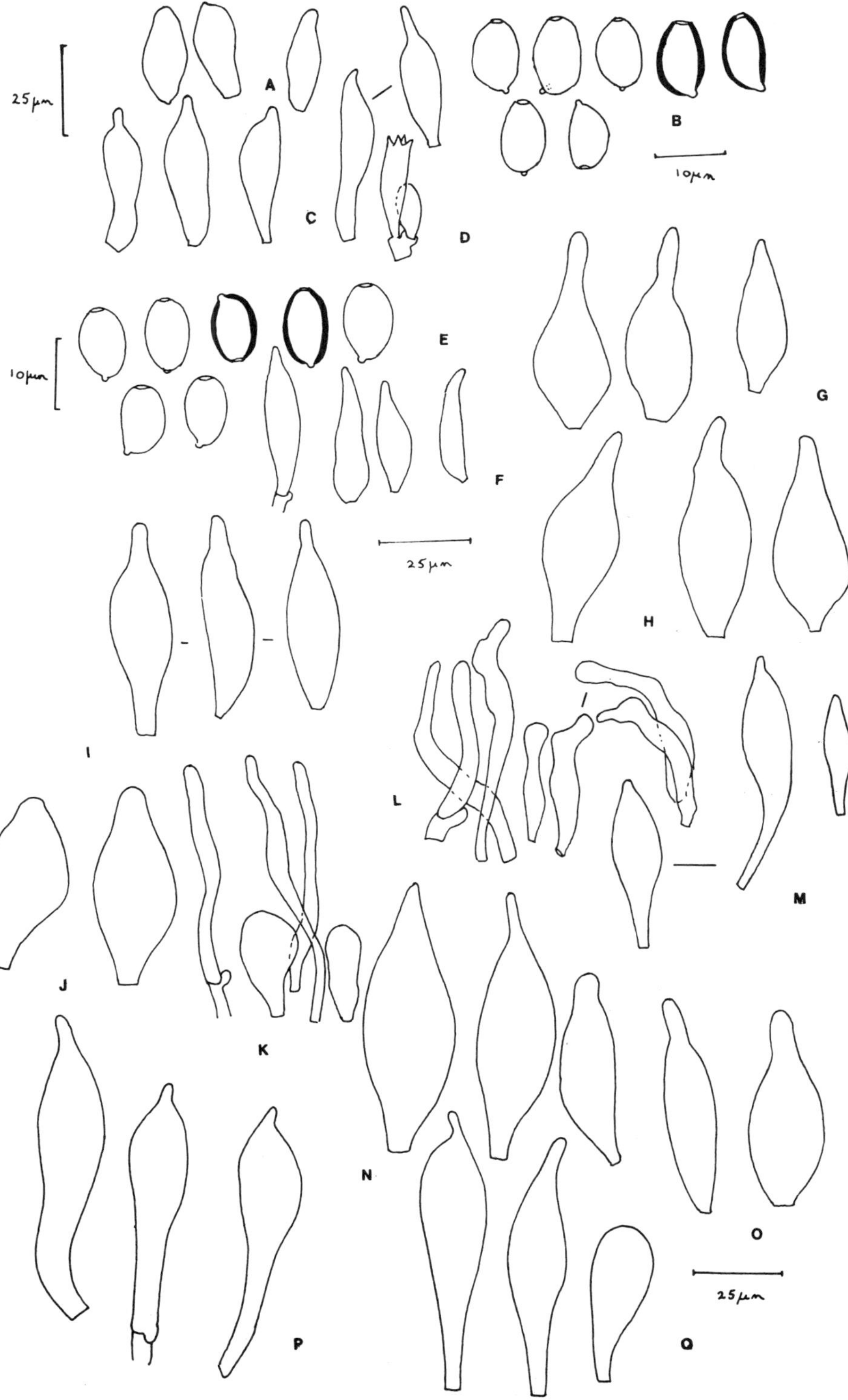

Fig. 7. AGROCYBE

<u>Agrocybe</u> <u>pediades</u>: A-D <u>Taylor</u> 1325, A. Caulocystidia;
B. Cheilocystidia; C. Basidium; D. Basidiospores.
E & G-I <u>Taylor</u> 1263, E. Caulocystidia,
G. Cheilocystidia; H. Basidia; I. Basidiospores.
F,J,K <u>Colenso</u> b113, F. Cheilocystidia; J. Cheilocystidia
of large basidioma; K. Cheilocystidia of small
basidioma.   L. Basidiospores <u>A.</u> <u>erebia</u> ex <u>K</u>.
M-N '<u>Agaricus</u> <u>strophosus</u>' ex <u>K</u>, M. Basidiospores;
N. Basidia.

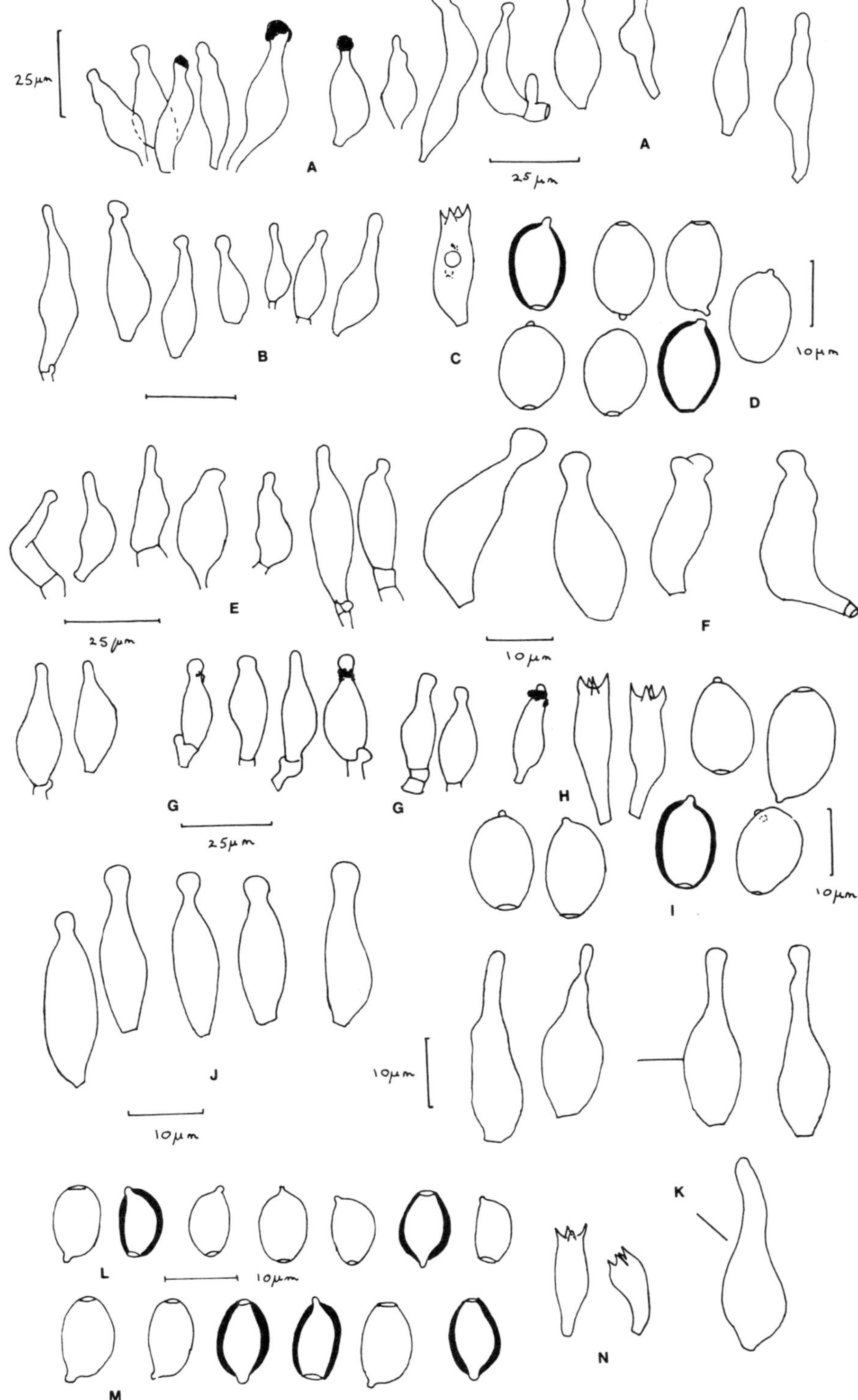

Fig. 8. AGROCYBE

Agrocybe praecox: A & I-J Taylor 318, A. Habit sketch
and section (A'); I. Pleurocystidia; J. Cheilocystidia.
B-D Taylor 317, B. Habit sketch and section (B');
C. Pleurocystidia; D. Cheilocystidia. E. Pleurocystidia
Orton 2102. F-G Orton 2101, F. Cheilocystidia;
G. Pleurocystidia. H. Pleurocystidia PDD 28569;
K. Pleurocystidia Orton 2103; L. Pleurocystidia PDD
29685. M-N PDD 24770, M. Pleurocystidia; N. Cheilocystidia.
O-P Austwick 1942, O. Cheilocystidia; P. Pleurocystidia.

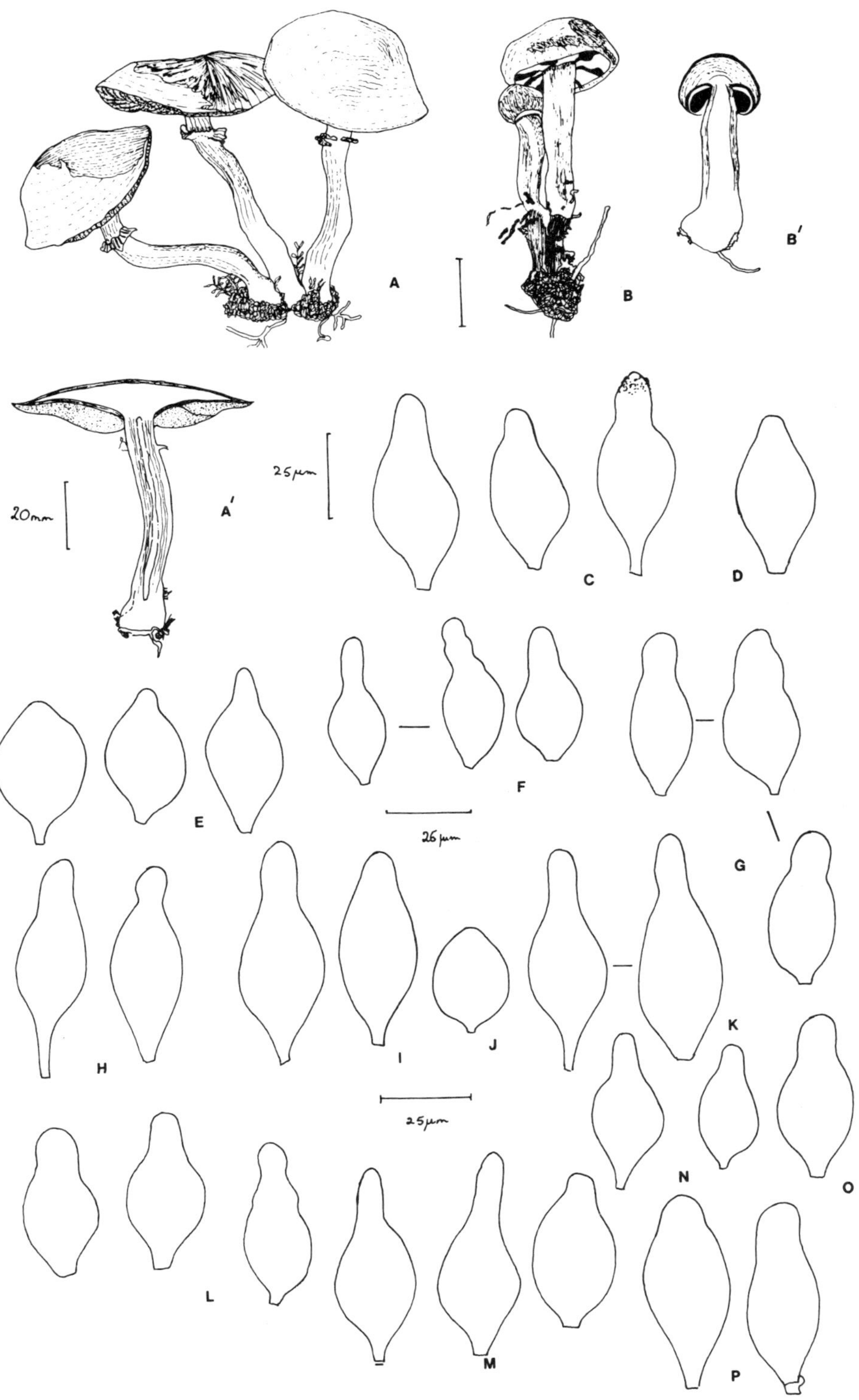

A
B
B'
A'
20mm
25μm
C
D
E
26μm
F
G
H
I
J
K
25μm
L
M
N
O
P

Fig. 9. AGROCYBE

Agrocybe praecox: A-D Segedin 751, A. Basidiospores;
B. Pleurocystidia; C & C' Caulocystidia; D. Pileipellis.
E-G Taylor 1254A, E. Basidiospores; F. Cheilocystidia;
G. Pleurocystidia. H-K Taylor 1261, H. Basidiospores;
I. Pleurocystidia; J. Basidia; K. Cheilocystidia.
L-O Austwick 1921, L. Basidiospores; M. Pleurocystidia;
N. Caulocystidia; O. Cheilocystidia; P-U Segedin 1719,
P. Basidiospores; Q. Pileipellis; R. Pleurocystidia;
S. Caulocystidia; T. Cheilocystidia; U. Basidia.

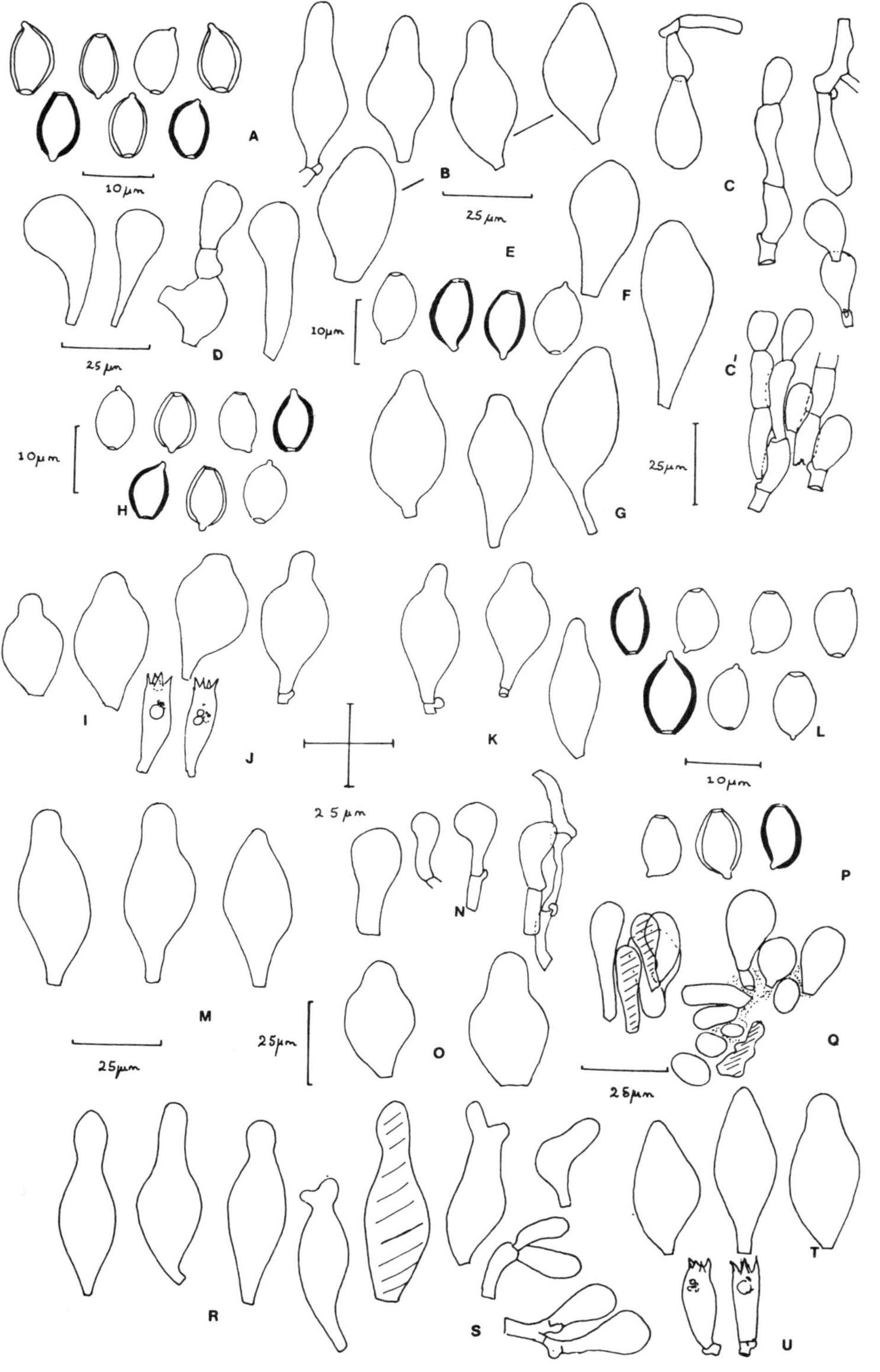

Fig. 10. BOLBITIUS

<u>Bolbitius</u> <u>muscicola</u>: A. Basidioma with section (B) <u>Taylor</u> 665.
C. Basidioma with section (D) <u>Taylor</u> 84.   E. Basidioma
with section (F) <u>Taylor</u> 342. G-I <u>Taylor</u> 665,
G. Basidioma with section (H); I. LS Pileipellis continuing
in I'. J-O <u>Taylor</u> 1322, J. Cheilocystidia; K. Basidia;
L. Caulocystidia; M. Basidiospores; N-O Pipeillis.

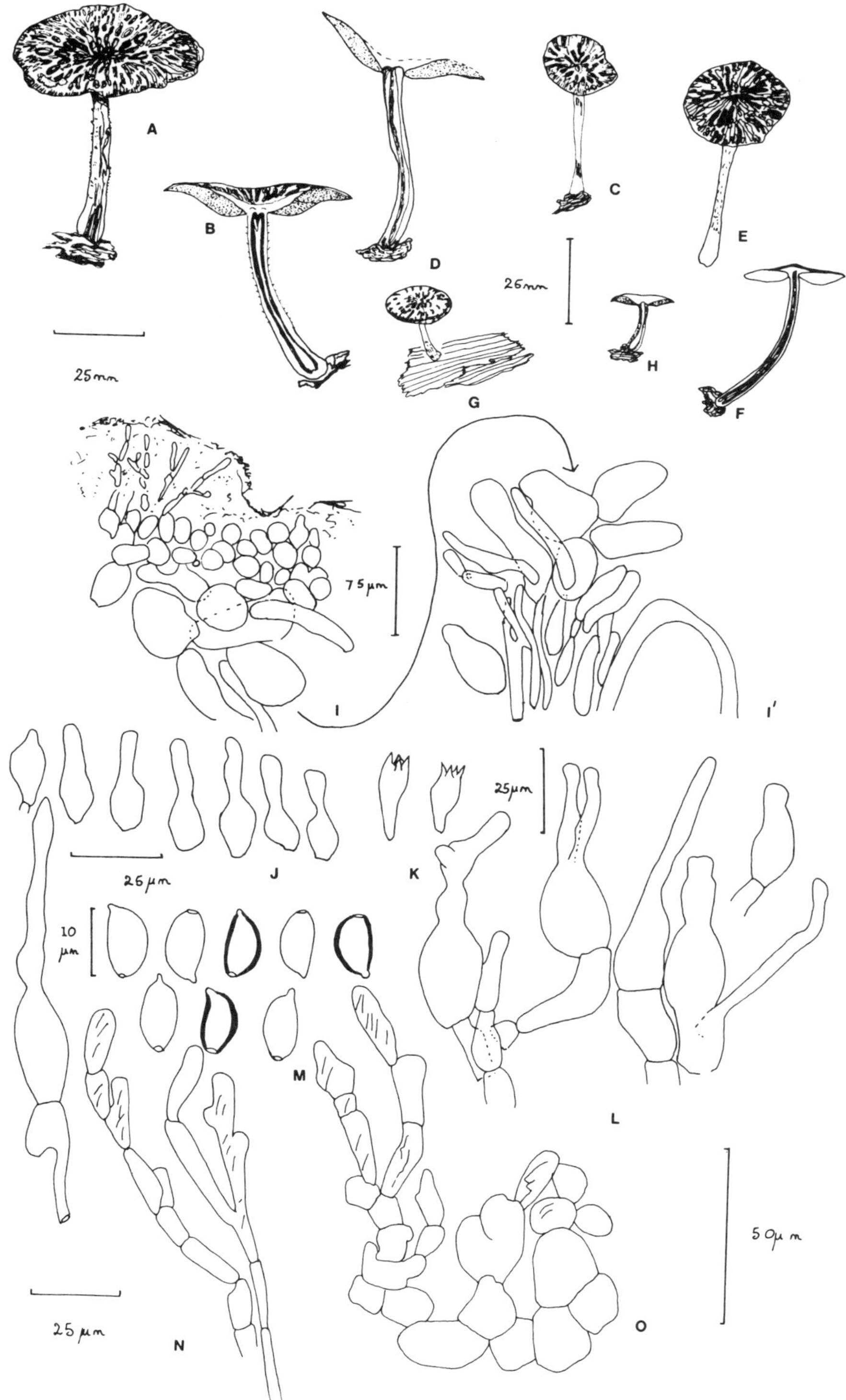

A
B
C
D
E
F
G
H
I
I'
J
K
L
M
N
O
25mm
25mm
25mm
75µm
25µm
25µm
10µm
50µm
25µm

Fig. 11. BOLBITIUS

<u>Bolbitius</u> <u>muscicola</u>: A & H <u>Stevenson</u> (holotype),
A. Cheilocystidia; H. Basidiospores, B-D <u>Horak</u> 67/112,
B. Basidiospores; C. Cheilocystidia; D. Caulocystidia.
E-G <u>Horak</u> 67/46, E. Cheilocystidia; F. Caulocystidia;
G. Basidiospores. I-M <u>Horak</u> 68/690, I. Cheilocystidia;
J. Basidia; K. Basidiospores; L. Caulocystidia; M.Pileipellis
units. N-P <u>Taylor</u> 319, N. Cheilocystidia; O-P Pileipellis.
Q-T <u>Taylor</u> 342, Q. Cheilocystidia; R. Pileipellis;
S. Caulocystidia; T. Basidium.

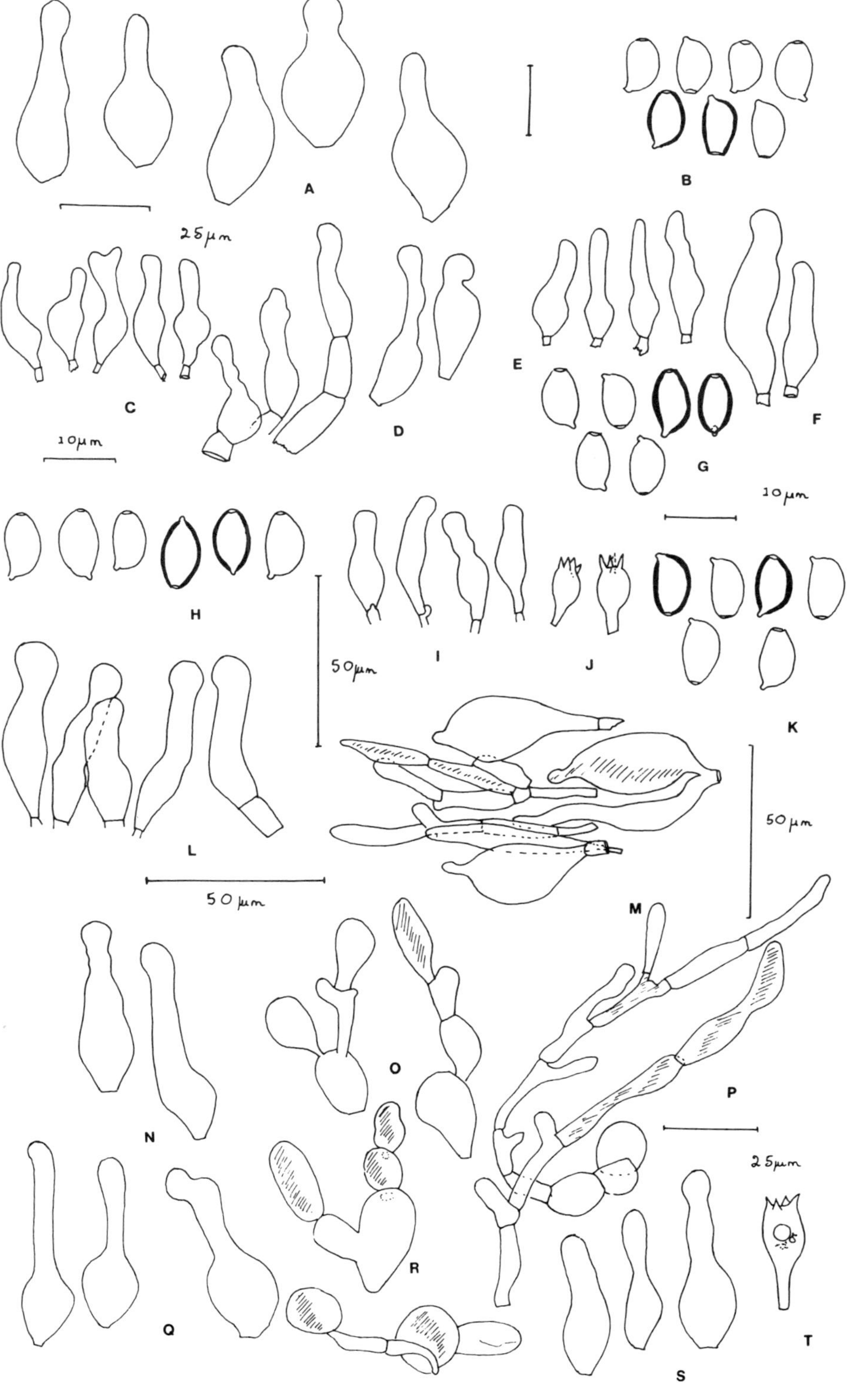

Fig. 12. BOLBITIUS

Bolbitius vitellinus: A-B Taylor 313, A. Basidiomata
with section (B); C-D & H Taylor 570, C. Basidiomata with
section (D); H. Cheilocystidia.
Bolbitius sp.2: Taylor 246, E. Basidiomata.
B. titubans: F-G Taylor 190, G. Immature basidiomata
with section (F).

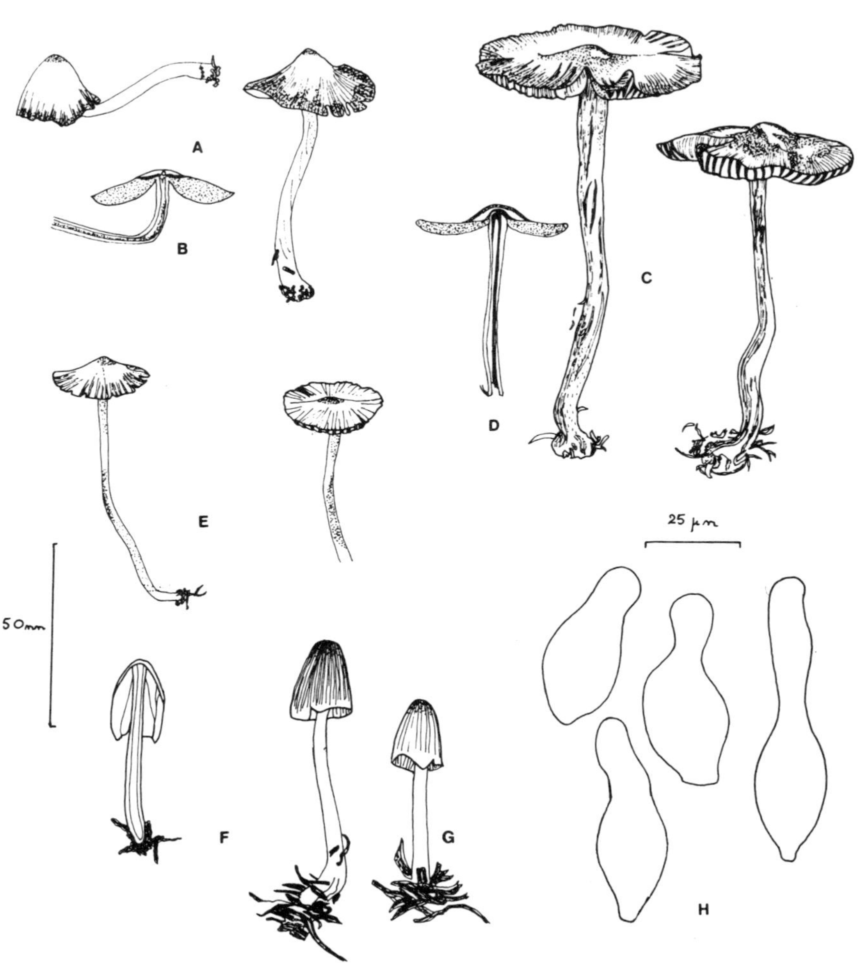

Fig. 13. BOLBITIUS AND CONOCYBE

B. titubans: A,D-F Segedin 1161, A.Basidiospores
(A' biporate); D. Pileipellis units; E. Pleurocystidia;
F. Caulocystidia. B-C Segedin 1171, B & B' Cheilocystidia;
C. Basidiospores. G-J Taylor 1260, G. Basidiospores
(G' biporate); H. Pileipellis; I. Caulocystidia;
J. Gill-margin.
Conocybe sp: Taylor 1254B, K. Basidiospores.

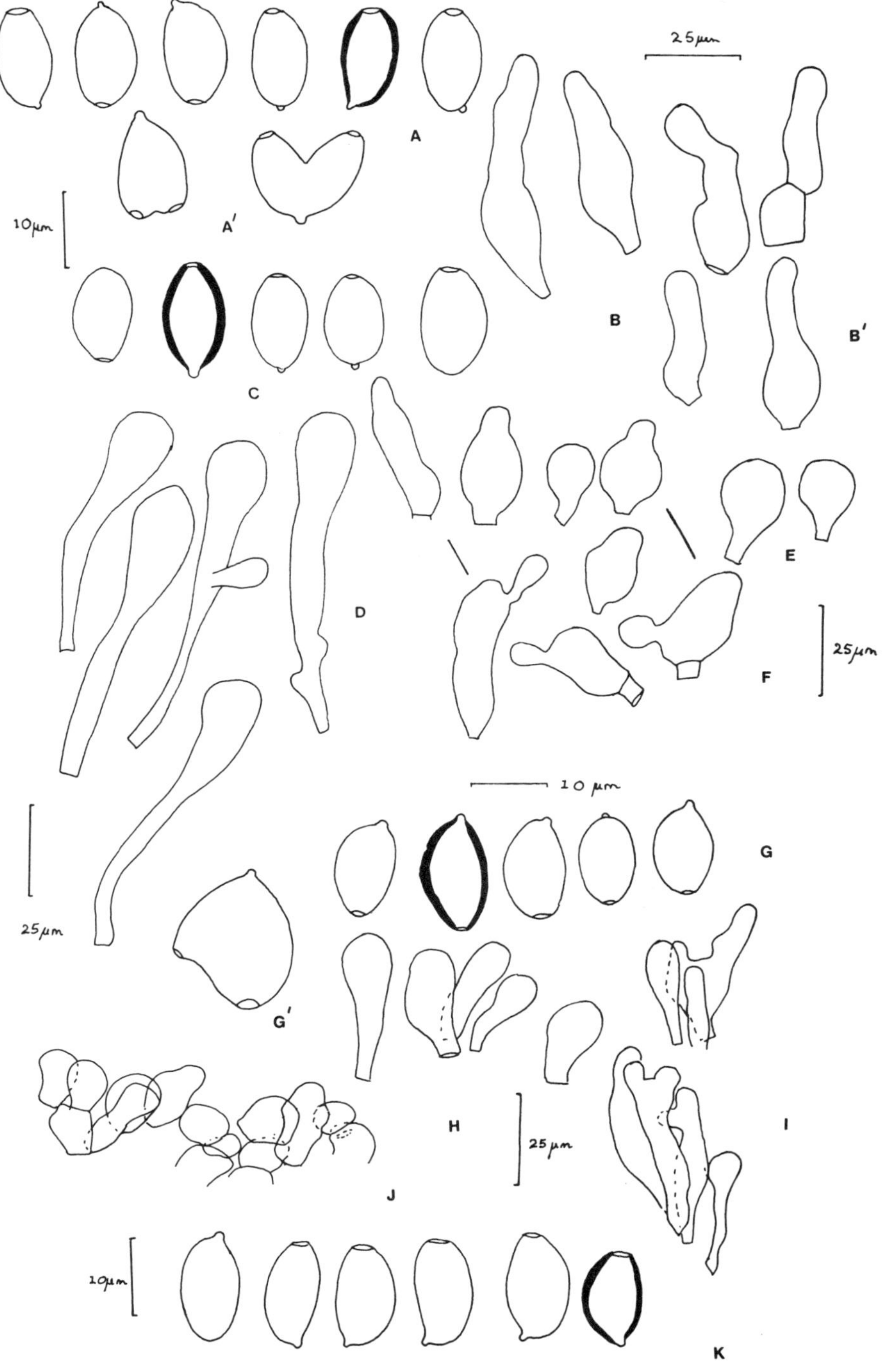

Fig. 14 CONOCYBE

Conocybe aff. vexans: A-E Horak 69/20, A. Basidiomata with section; B. Caulocystidia; C. Cheilocystidia; D. Basidia; E. Basidiospores. F-I Horak 68/440, F. Basidiomata; G. Cheilocystidia; H. Basidiospores; I. Pileipellis.
C. rugosa: J-N Taylor 1280, J. Basidioma and section; K. Cheilocystidia; L. Basidia; M. Basidiospores; N. Caulocystidia.
C. gracilenta; Horak 68/303 (holotype), O. Basidiomata; P. Caulocystidia; Q. Cheilocystidia; R. Basidiospores.

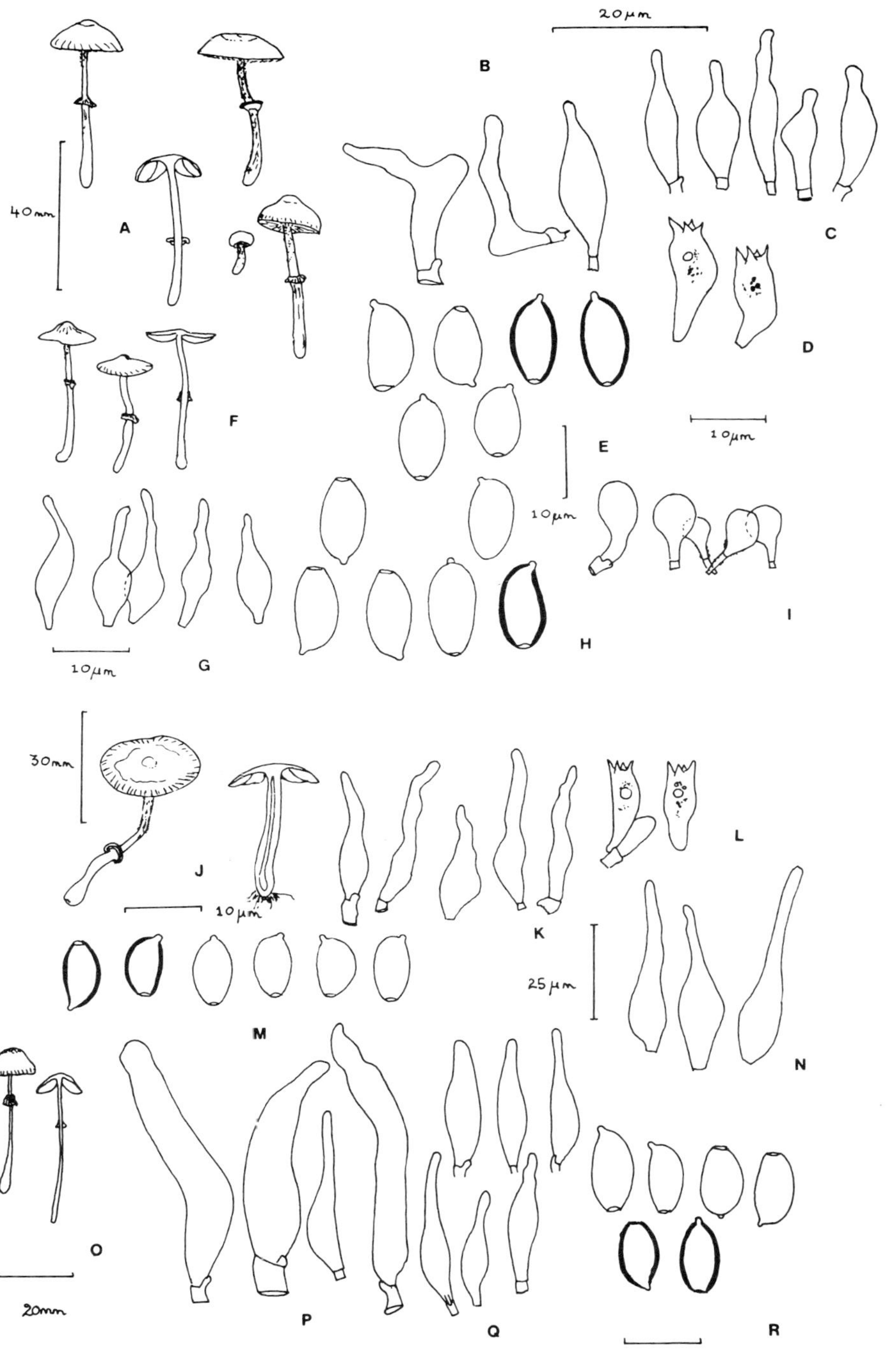

Fig. 15. CONOCYBE

<u>Conocybe piloselloides</u>: A, C-F <u>Taylor</u> 1055, A. Habit
sketch and section; C. Hairs on stipe; D. Cheilocystidia;
E. Pileipellis units; F. Blematogen fragments.
<u>C.</u> <u>novaezelandiae</u>: B, G-H <u>Taylor</u> 178 (holotype),
B. Habit sketch and section; G. Cheilocystidia;
H. Caulocystidia.
<u>C.</u> <u>mesospora</u>; I-M <u>Horak</u> 61/67,
I. Basidiomata with section; J. Caulocystidia;
K. Basidiospores; L. Cheilocystidia; M. Basidia.

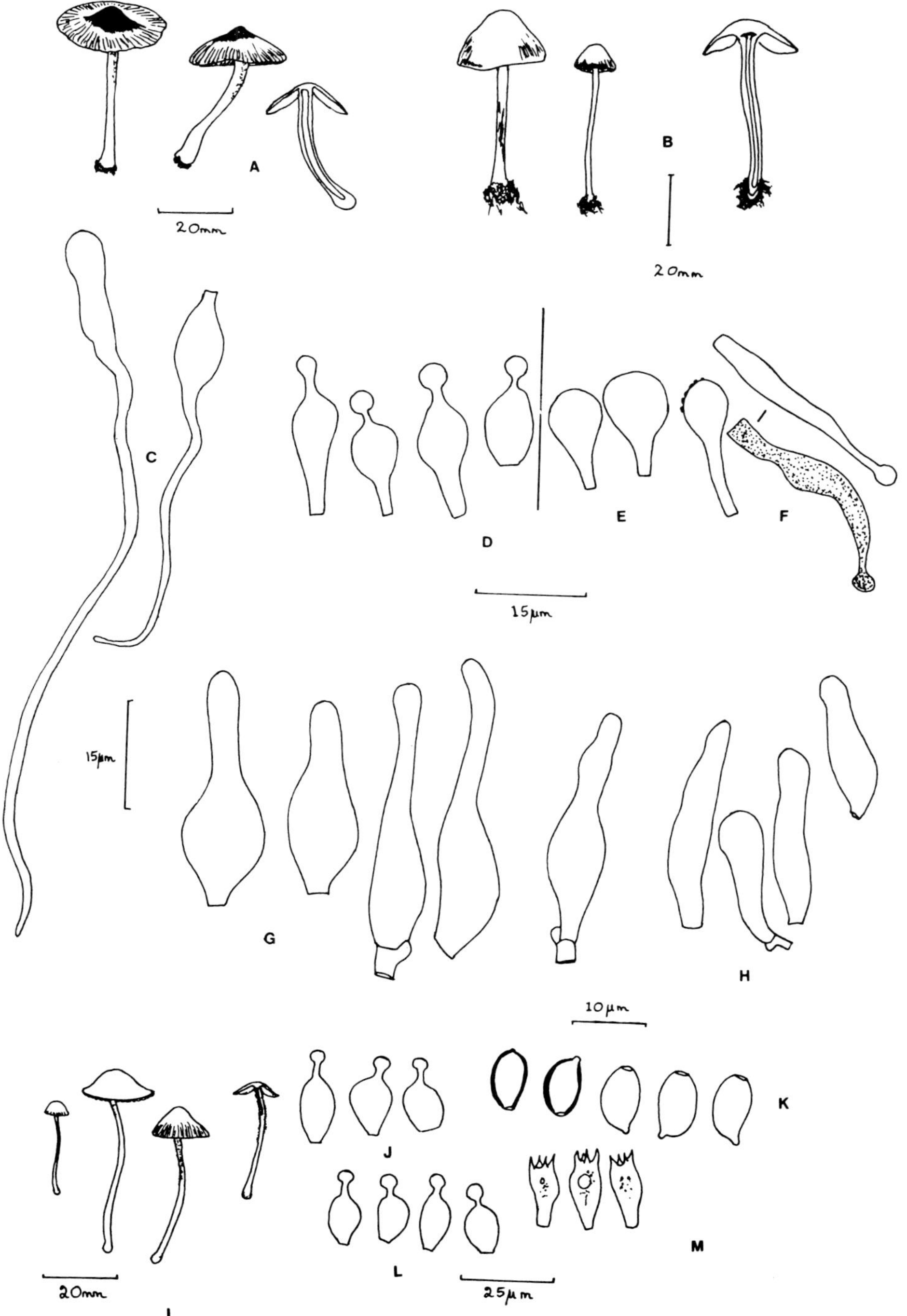

Fig. 16. CONOCYBE

Conocybe sp. 1: A, R-S Taylor 64, A. Habit sketch and
section; R. Basidiospores; S. Cheilocystidia.
Conocybe sp. 4: B-D Segedin 165, B. Basidiospores;
C. Cheilocystidia; D. Pileipellis units.
Conocybe pubescens: E-I Taylor 1346, E. Pileipellis
units; F. Basidia; G. Basidiospores; H. Cheilocystidia;
I. Caulocystidia; I'. Hair from caulohymenium.
Conocybe sp. 3: J-M, P-Q Segedin 219, J. Basidiospores;
K-L Cheilocystidia; M. Basidia; P-Q Pileipellis units.
C. huijsmanii: N & O Taylor 1151 (AK 1),
T & V Taylor 1151 (AK 2), N & V Basidiospores;
O. Cheilocystidia; T. Basidia.
C. mesospora: U, W-X Horak 68/154,
U. Basidiospores; W. Cheilocystidia; X. Pileipellis.

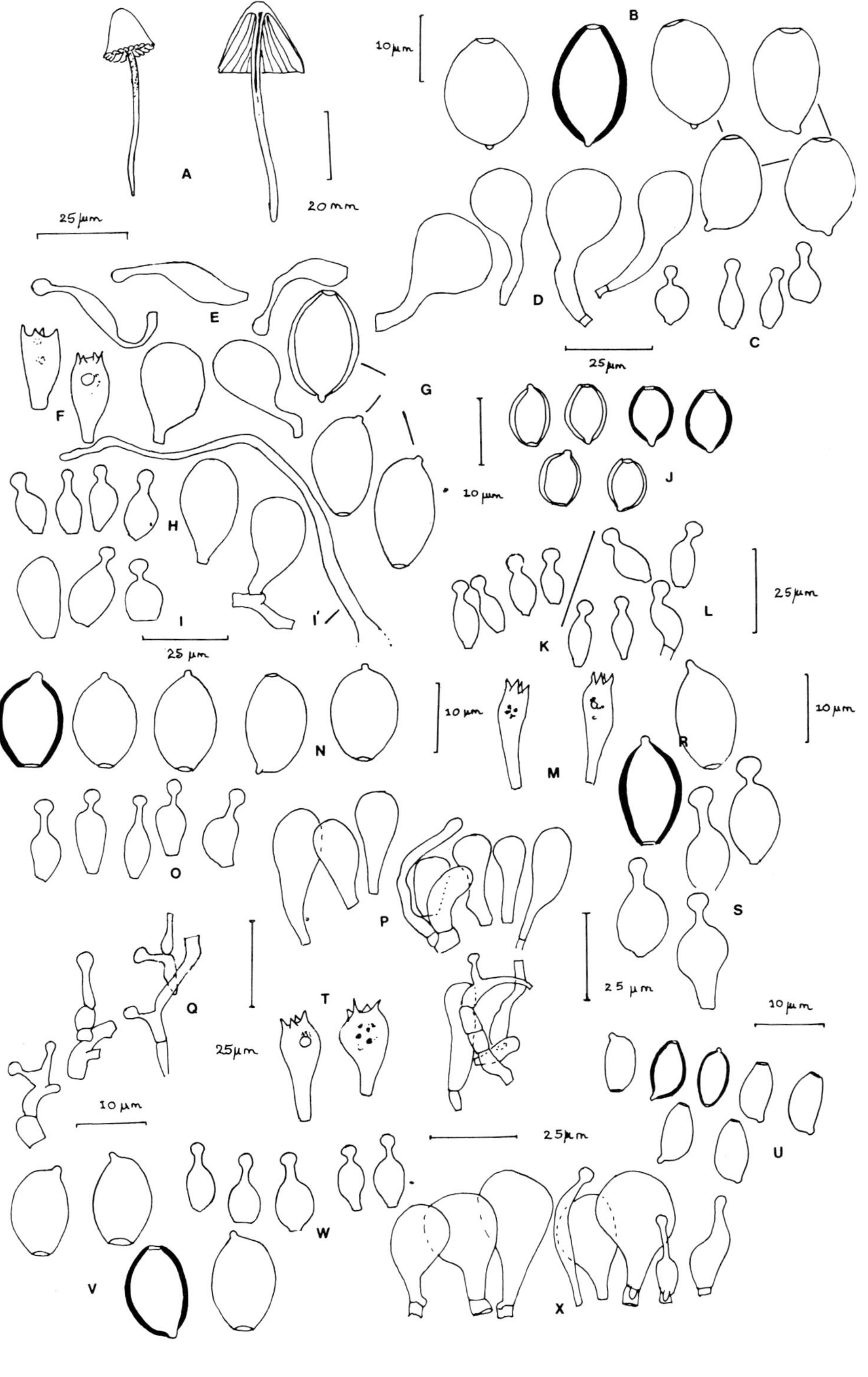

<u>Plate</u> <u>1.</u> Basidiospores: A-C <u>C.</u> <u>horakii</u>, holotype.
D. <u>C.</u> <u>dumetorum</u>.

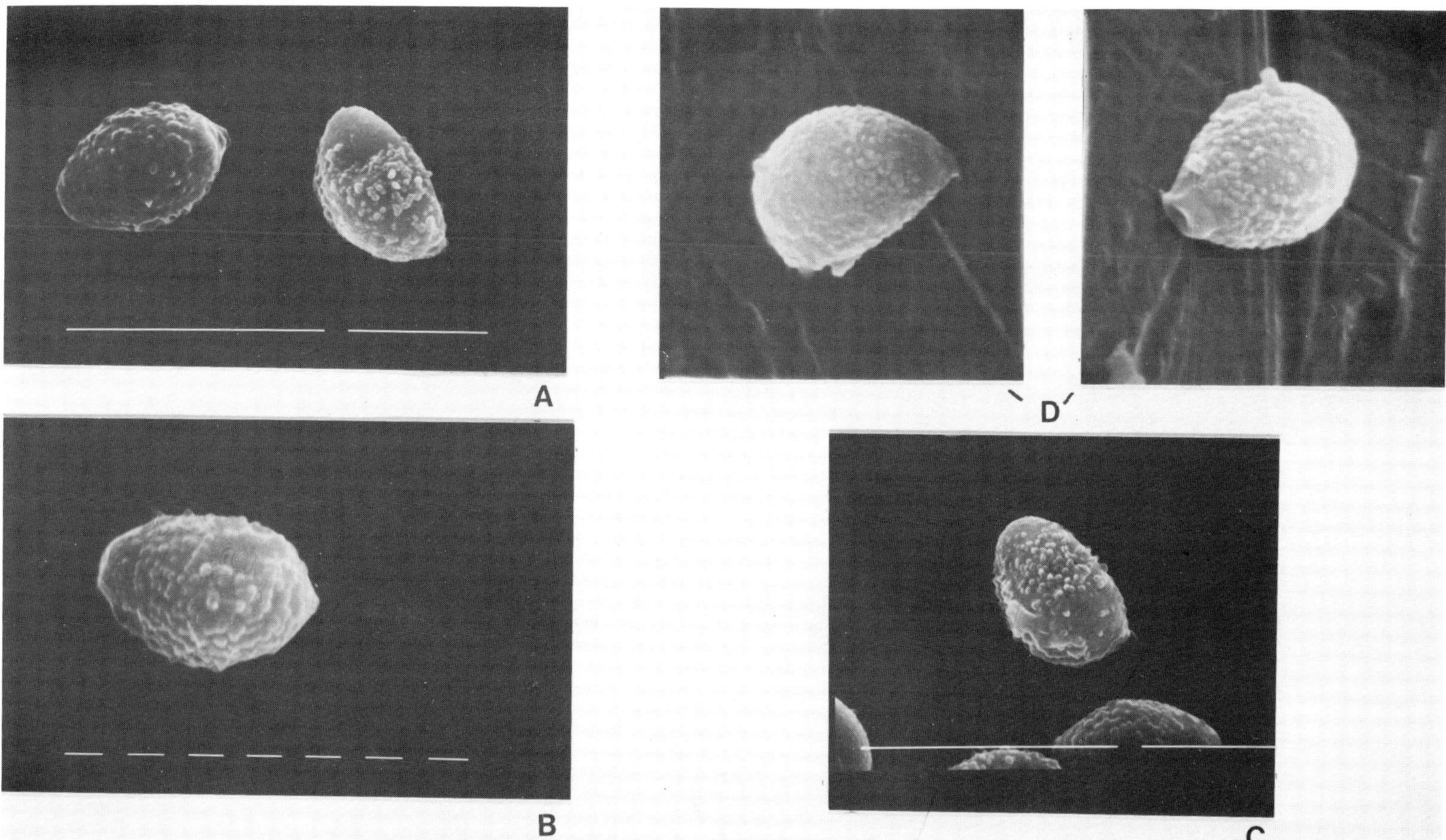

A
B
C
D´